2D Nanostructures for Optoelectronic and Green Energy Devices

2D Nanostructures for Optoelectronic and Green Energy Devices

Editors

Jung Inn Sohn
Sangyeon Pak

MDPI • Basel • Beijing • Wuhan • Barcelona • Belgrade • Manchester • Tokyo • Cluj • Tianjin

Editors
Jung Inn Sohn
Dongguk University-Seoul
Republic of Korea

Sangyeon Pak
Hongik University
Republic of Korea

Editorial Office
MDPI
St. Alban-Anlage 66
4052 Basel, Switzerland

This is a reprint of articles from the Special Issue published online in the open access journal *Nanomaterials* (ISSN 2079-4991) (available at: https://www.mdpi.com/journal/nanomaterials/special_issues/nano_optoelectro_energy).

For citation purposes, cite each article independently as indicated on the article page online and as indicated below:

LastName, A.A.; LastName, B.B.; LastName, C.C. Article Title. *Journal Name* **Year**, *Volume Number*, Page Range.

ISBN 978-3-0365-7258-1 (Hbk)
ISBN 978-3-0365-7259-8 (PDF)

© 2023 by the authors. Articles in this book are Open Access and distributed under the Creative Commons Attribution (CC BY) license, which allows users to download, copy and build upon published articles, as long as the author and publisher are properly credited, which ensures maximum dissemination and a wider impact of our publications.

The book as a whole is distributed by MDPI under the terms and conditions of the Creative Commons license CC BY-NC-ND.

Contents

About the Editors vii

Sangyeon Pak and Jung Inn Sohn
2D Nanostructures for Optoelectronic and Green Energy Devices
Reprinted from: *Nanomaterials* **2023**, *13*, 1070, doi:10.3390/nano13061070 1

A-Rang Jang
Tuning Schottky Barrier of Single-Layer MoS₂ Field-Effect Transistors with Graphene Electrodes
Reprinted from: *Nanomaterials* **2022**, *12*, 3038, doi:10.3390/nano12173038 3

Sangyeon Pak
Controlled p-Type Doping of MoS₂ Monolayer by Copper Chloride
Reprinted from: *Nanomaterials* **2022**, *12*, 2893, doi:10.3390/nano12172893 15

Ki Hoon Shin, Min-Kyu Seo, Sangyeon Pak, A-Rang Jang and Jung Inn Sohn
Observation of Strong Interlayer Couplings in WS₂/MoS₂ Heterostructures via Low-Frequency Raman Spectroscopy
Reprinted from: *Nanomaterials* **2022**, *12*, 1393, doi:10.3390/nano12091393 23

Konthoujam James Singh, Hao-Hsuan Ciou, Ya-Hui Chang, Yen-Shou Lin, Hsiang-Ting Lin, Po-Cheng Tsai, Shih-Yen Lin, et al.
Optical Mode Tuning of Monolayer Tungsten Diselenide (WSe₂) by Integrating with One-Dimensional Photonic Crystal through Exciton–Photon Coupling
Reprinted from: *Nanomaterials* **2022**, *12*, 425, doi:10.3390/nano12030425 33

Jong Chan Yoon, Zonghoon Lee and Gyeong Hee Ryu
Atomic Arrangements of Graphene-like ZnO
Reprinted from: *Nanomaterials* **2021**, *11*, 1833, doi:10.3390/nano11071833 47

Sijia Miao, Tianle Liu, Yujian Du, Xinyi Zhou, Jingnan Gao, Yichu Xie, Fengyi Shen, et al.
2D Material and Perovskite Heterostructure for Optoelectronic Applications
Reprinted from: *Nanomaterials* **2022**, *12*, 2100, doi:10.3390/nano12122100 55

Puran Pandey, Min-Kyu Seo, Ki Hoon Shin, Young-Woo Lee and Jung Inn Sohn
Hierarchically Assembled Plasmonic Metal-Dielectric-Metal Hybrid Nano-Architectures for High-Sensitivity SERS Detection
Reprinted from: *Nanomaterials* **2022**, *12*, 401, doi:10.3390/nano12030401 89

Sung Min Wi, Jihong Kim, Suok Lee, Yu-Rim Choi, Sung Hoon Kim, Jong Bae Park, Younghyun Cho, et al.
A Redox-Mediator-Integrated Flexible Micro-Supercapacitor with Improved Energy Storage Capability and Suppressed Self-Discharge Rate
Reprinted from: *Nanomaterials* **2021**, *11*, 3027, doi:10.3390/nano11113027 99

Xianyu Liu, Liwen Ma, Yehong Du, Qiongqiong Lu, Aikai Yang and Xinyu Wang
Vanadium Pentoxide Nanofibers/Carbon Nanotubes Hybrid Film for High-Performance Aqueous Zinc-Ion Batteries
Reprinted from: *Nanomaterials* **2021**, *11*, 1054, doi:10.3390/nano11041054 111

Luomeng Chao, Changwei Sun, Jiaxin Li, Miao Sun, Jia Liu and Yonghong Ma
Transparent Heat Shielding Properties of Core-Shell Structured Nanocrystalline $Cs_xWO_3@TiO_2$
Reprinted from: *Nanomaterials* **2022**, *12*, 2806, doi:10.3390/nano12162806 121

About the Editors

Jung Inn Sohn

Jung Inn Sohn is currently a professor in the Division of Physics and Semiconductor Science at Dongguk University, Korea. His research group is developing new low-dimensional materials and exploring their fundamental physical properties and new functions for potential applications in energy and optoelectronics. He was formerly an academic faculty member at the University of Oxford and a senior researcher at Samsung Advanced Institute of Technology and the University of Cambridge. His research is described in more than 150 SCI articles in top journals including *Nature, Nature Communications, Energy & Environmental Science*, as well as 18 cover picture articles and 4 review articles. He has an h-index of 41 and 26 patents to his name.

Sangyeon Pak

Sangyeon Pak is currently an assistant professor in the School of Electronic and Electrical Engineering at Hongik University, Korea. His research focuses on developing semiconductor devices based on two-dimensional materials. He received his bachelor's degree in Electrical Engineering at the University of Wisconsin-Madison in 2014 and DPhil degree in the Department of Engineering Science at University of Oxford in 2019. He worked as a postdoc researcher at the Institute of Basic Science and Department of Physics at Sungkyunkwan University (2019–2022).

Editorial

2D Nanostructures for Optoelectronic and Green Energy Devices

Sangyeon Pak [1],* and Jung-Inn Sohn [2],*

1 School of Electronic and Electrical Engineering, Hongik University, Seoul 04066, Republic of Korea
2 Division of Physics and Semiconductor Science, Dongguk University-Seoul, Seoul 04620, Republic of Korea
* Correspondence: spak@hongik.ac.kr (S.P.); junginn.sohn@dongguk.edu (J.-I.S.)

Two-dimensional (2D) materials and nanostructures have gathered significant attention due to their excellent mechanical properties [1], unique electrical and optical characteristics [2,3], large surface-to-volume ratio, and chemical and environmental stability [4]. These features have led to the discovery of a large new family of 2D materials and a vast range of possible applications ranging from optoelectronics and electronics to energy conversion and saving applications. With the extensive catalogue of available 2D materials, ranging from metallic layers to semiconductors and insulators, their applications hold great promise for future innovative research in science and technology.

In recent years, many colleagues have been developing fundamental studies dedicated specifically to the design of 2D materials in optoelectronic and green energy devices due to their excellent opto-electrical and electrochemical performance. These research studies sit at the interface between engineering, material science, physics, and chemistry, and the importance of 2D materials and nanostructures in such applications calls for intensive experimental research assisted with engineering their fundamental and interface properties for enhancing the performance of various devices.

In this Special Issue "2D Nanostructures for Optoelectronic and Green Energy Devices", we have collected nine high-quality, original research papers and one comprehensive review paper by outstanding scientists and engineers from relevant fields, covering the topics in optical properties and couplings in 2D nanostructures, [5–8] spectroscopic analysis in atomic scale, [9] 2D transistors, [10,11] 2D optoelectronics, [12] and 2D energy storage applications [13,14].

The synthesis and characterization of the new fundamental properties of 2D nanostructures are of fundamental importance in order to utilize the 2D nanostructures for optoelectronic and green energy device applications. Shin et al. [6] reported interlayer coupling effects in 2D heterostructures using low-frequency Raman spectroscopy, offering a route to observe the quality of the interface in 2D nanostructures. Singh et al. [7] and Pandey et al. [8] reported optical coupling in nanostructures and substrates, which can be used in nanophotonics and detection applications. Yoon et al. [9] reported the atomic arrangement of graphene-like ZnO examined using transmission electron spectroscopy. A-Rang Jang [11] reported graphene contact in a 2D transistor, which was effective in lowering the Schottky barrier in metal-2D semiconductor contact. Sangyeon Pak [10] reported a simple and effective fabrication route of p-type doping 2D MoS_2 simply by spin coating $CuCl_2$ molecules.

Two-dimensional nanostructures are also found extensively in the field of energy storage applications where nanomaterials can be used as the effective electrode materials with a large surface area in batteries and supercapacitors. The work by Wi and coworkers [13] reported layered graphite structures fabricated using the direct laser scribing of PI substrate, which is an effective electrode material in micro-supercapacitors when integrated with hydroquinone gel electrolyte. Such flexible micro-supercapacitors promise future energy storage components, especially in wearable applications. Liu et al. [14] utilized

Citation: Pak, S.; Sohn, J.-I. 2D Nanostructures for Optoelectronic and Green Energy Devices. *Nanomaterials* **2023**, *13*, 1070. https://doi.org/10.3390/nano13061070

Received: 27 February 2023
Revised: 14 March 2023
Accepted: 14 March 2023
Published: 16 March 2023

Copyright: © 2023 by the authors. Licensee MDPI, Basel, Switzerland. This article is an open access article distributed under the terms and conditions of the Creative Commons Attribution (CC BY) license (https://creativecommons.org/licenses/by/4.0/).

vanadium pentoxide nanofiber/carbon nanotube hybrid films as binder-free cathodes for zinc-ion batteries, achieving a high energy density and stable cyclability, demonstrating the promises for large-scale energy storage applications.

Last, the Special Issue includes a comprehensive review on the applications on 2D materials, perovskites, and 2D material/perovskite heterostructures for applications to optoelectronic devices, including solar cells and photodetectors, nicely summarized by Yuljae Cho and his co-workers [12]. Especially, the review summarizes various synthetic methods for 2D materials and perovskite materials, and the photodetector performance of various materials was adequately compared and summarized.

To summarize, this Special Issue is expected to attract and enrich readers through featuring all of the above-mentioned research articles and review articles. Especially, we express our sincere thanks to all the authors, reviewers, and editors that made a contribution to this Special Issue.

Funding: This work was supported by the National Research Foundation of Korea (NRF) grant funded by the Korean government (MSIT) (Grant no. 2022R1F1A1063997) and 2023 Hongik University Research Fund.

Data Availability Statement: Not Applicable.

Conflicts of Interest: The authors declare no conflict of interest.

References

1. Pak, S.; Lee, J.; Jang, A.R.; Kim, S.; Park, K.-H.; Sohn, J.I.; Cha, S. Strain-Engineering of Contact Energy Barriers and Photoresponse Behaviors in Monolayer MoS2 Flexible Devices. *Adv. Funct. Mater.* **2020**, *30*, 2002023. [CrossRef]
2. Kim, T.; Lim, J.; Byeon, J.; Cho, Y.; Kim, W.; Hong, J.; Jin Heo, S.; Eun Jang, J.; Kim, B.-S.; Hong, J.; et al. Electronic Modulation of Semimetallic Electrode for 2D van der Waals Devices. *Small Struct.* **2023**, 2200274. [CrossRef]
3. Kim, T.; Pak, S.; Lim, J.; Hwang, J.S.; Park, K.H.; Kim, B.S.; Cha, S. Electromagnetic Interference Shielding with 2D Copper Sulfide. *ACS Appl. Mater. Interfaces* **2022**, *14*, 13499–13505. [CrossRef] [PubMed]
4. Lim, J.; Kim, T.; Byeon, J.; Park, K.-H.; Hong, J.; Pak, S.; Cha, S. Energy level modulation of MoS2 monolayers by halide doping for an enhanced hydrogen evolution reaction. *J. Mater. Chem. A* **2022**, *10*, 23274–23281. [CrossRef]
5. Chao, L.; Sun, C.; Li, J.; Sun, M.; Liu, J.; Ma, Y. Transparent Heat Shielding Properties of Core-Shell Structured Nanocrystalline Cs(x)WO(3)@TiO(2). *Nanomaterials* **2022**, *12*, 2806. [CrossRef] [PubMed]
6. Shin, K.H.; Seo, M.K.; Pak, S.; Jang, A.R.; Sohn, J.I. Observation of Strong Interlayer Couplings in WS(2)/MoS(2) Heterostructures via Low-Frequency Raman Spectroscopy. *Nanomaterials* **2022**, *12*, 1393. [CrossRef] [PubMed]
7. James Singh, K.; Ciou, H.H.; Chang, Y.H.; Lin, Y.S.; Lin, H.T.; Tsai, P.C.; Lin, S.Y.; Shih, M.H.; Kuo, H.C. Optical Mode Tuning of Monolayer Tungsten Diselenide (WSe(2)) by Integrating with One-Dimensional Photonic Crystal through Exciton-Photon Coupling. *Nanomaterials* **2022**, *12*, 425. [CrossRef] [PubMed]
8. Pandey, P.; Seo, M.K.; Shin, K.H.; Lee, Y.W.; Sohn, J.I. Hierarchically Assembled Plasmonic Metal-Dielectric-Metal Hybrid Nano-Architectures for High-Sensitivity SERS Detection. *Nanomaterials* **2022**, *12*, 401. [CrossRef] [PubMed]
9. Yoon, J.C.; Lee, Z.; Ryu, G.H. Atomic Arrangements of Graphene-like ZnO. *Nanomaterials* **2021**, *11*, 1833. [CrossRef] [PubMed]
10. Pak, S. Controlled p-Type Doping of MoS2 Monolayer by Copper Chloride. *Nanomaterials* **2022**, *12*, 2893. [CrossRef] [PubMed]
11. Jang, A.R. Tuning Schottky Barrier of Single-Layer MoS(2) Field-Effect Transistors with Graphene Electrodes. *Nanomaterials* **2022**, *12*, 3038. [CrossRef] [PubMed]
12. Miao, S.; Liu, T.; Du, Y.; Zhou, X.; Gao, J.; Xie, Y.; Shen, F.; Liu, Y.; Cho, Y. 2D Material and Perovskite Heterostructure for Optoelectronic Applications. *Nanomaterials* **2022**, *12*, 1489. [CrossRef] [PubMed]
13. Wi, S.M.; Kim, J.; Lee, S.; Choi, Y.R.; Kim, S.H.; Park, J.B.; Cho, Y.; Ahn, W.; Jang, A.R.; Hong, J.; et al. A Redox-Mediator-Integrated Flexible Micro-Supercapacitor with Improved Energy Storage Capability and Suppressed Self-Discharge Rate. *Nanomaterials* **2021**, *11*, 3027. [CrossRef] [PubMed]
14. Liu, X.; Ma, L.; Du, Y.; Lu, Q.; Yang, A.; Wang, X. Vanadium Pentoxide Nanofibers/Carbon Nanotubes Hybrid Film for High-Performance Aqueous Zinc-Ion Batteries. *Nanomaterials* **2021**, *11*, 1054. [CrossRef] [PubMed]

Disclaimer/Publisher's Note: The statements, opinions and data contained in all publications are solely those of the individual author(s) and contributor(s) and not of MDPI and/or the editor(s). MDPI and/or the editor(s) disclaim responsibility for any injury to people or property resulting from any ideas, methods, instructions or products referred to in the content.

Article

Tuning Schottky Barrier of Single-Layer MoS$_2$ Field-Effect Transistors with Graphene Electrodes

A-Rang Jang

Division of Electrical, Electronic and Control Engineering, Kongju National University, Cheonan 31080, Korea; arjang@kongju.ac.kr

Abstract: Two–dimensional materials have the potential to be applied in flexible and transparent electronics. In this study, single-layer MoS$_2$ field-effect transistors (FETs) with Au/Ti–graphene heteroelectrodes were fabricated to examine the effect of the electrodes on the electrical properties of the MoS$_2$ FETs. The contact barrier potential was tuned using an electric field. Asymmetrical gate behavior was observed owing to the difference between the MoS$_2$ FETs, specifically between the MoS$_2$ FETs with Au/Ti electrodes and those with graphene electrodes. The contact barrier of the MoS$_2$ FETs with Au/Ti electrodes did not change with the electric field. However, the contact barrier at the MoS$_2$–graphene interface could be modulated. The MoS$_2$ FETs with Au/Ti–graphene electrodes exhibited enhanced on/off ratios (~10^2 times) and electron mobility (~2.5 times) compared to the MoS$_2$ FETs with Au/Ti electrodes. These results could improve the understanding of desirable contact formation for high-performance MoS$_2$ FETs and provide a facile route for viable electronic applications.

Keywords: graphene; molybdenum disulfide; Schottky barrier

1. Introduction

Two–dimensional (2D) materials have attracted significant attention as potential candidates for next-generation electronics [1,2]. Graphene is considered to be one of the most promising 2D materials because of its unique electrical, mechanical, and optical properties. However, the widespread use of graphene in viable electronic device applications is limited by its zero-bandgap property, which considerably decreases the on/off ratio [3–6]. To overcome this limitation, graphene nanoribbons [7,8], bilayer graphene [9–12], and modified device architectures, such as vertical tunneling transistors, have been developed [13,14]. Although these devices have improved the on/off ratio, other desirable properties, such as mobility and current density, have deteriorated. Thus, there is an urgent requirement for 2D materials, including transition metal dichalcogenides (TMDs), with an appropriate bandgap and reasonable mobility to replace graphene. MoS$_2$ is one of the most promising TMDs because its bandgap is 1.3–1.8 eV depending on the number of layers. Single-layer MoS$_2$ films have a direct bandgap of 1.8 eV, whereas multilayer MoS$_2$ films have an indirect bandgap of 1.2 eV [15–17]. Owing to these unique properties, MoS$_2$ has been intensively studied for electronic and optoelectronic applications. In recent years, it has become possible to synthesize large-area single-layer MoS$_2$ via chemical vapor deposition (CVD) [18–22]. This has provided a major opportunity for next-generation electronic device applications. However, the contact barrier issue must be studied for electronic device applications of 2D materials [23,24]. Moreover, the performance of MoS$_2$ field-effect transistors (FETs) is lower than the theoretically predicted performance [25]. This discrepancy has been explained on the basis of charged impurities and localized states in MoS$_2$ [26–28]. Dominant scattering processes decrease carrier mobility. In addition, the contact at MoS$_2$–metal electrode interfaces is a critical issue. A tunneling barrier that is formed at the interface of a metal contact in an MoS$_2$ device [29] significantly reduces carrier mobility in single–layer MoS$_2$. This is one of the main reasons for the poor performance of single-layer MoS$_2$ FETs.

Sulfur atoms mediate the hybridization between a contact metal and Mo atoms, resulting in the tuning of the bandgap [30]. Furthermore, the bandgap of single-layer MoS_2 can be determined by the strength of the Mo–S covalent bonding [31]. Therefore, a systematic study of the effects of electrode materials on the performance of MoS_2 FETs can help resolve this critical issue and find a reliable method of improving the electrical properties of MoS_2 FETs. A charge accumulation region forms at metal–MoS_2 interfaces when a metal contact is used. This generally leads to the formation of an interface electric dipole, which modifies the interface band alignment [30]. This results in poor contact and an unexpected contact barrier between the metal and MoS_2. Owing to the challenges associated with metal electrodes, graphene has been considered as a suitable electrode material for MoS_2 FETs. Graphene and single-layer MoS_2 bond via van der Waals (vdW) forces, thereby creating a pristine interface. Furthermore, the contact barrier between graphene and MoS_2 can be controlled by tuning the work function of graphene (4.5 eV), which is quite similar to that of MoS_2 [32]. As the work function of graphene can be readily tuned by applying an electric field, graphene-based heterostructures have recently been studied in electronic devices [33–36]. For instance, the Schottky barrier formed between graphene and silicon can be tuned by approximately 200 meV as a function of the gate voltage [13]. Therefore, the contact barrier between graphene and MoS_2 can be tuned by applying an electric field.

Herein, we report high-performance single–layer MoS_2 FETs with graphene electrodes that exhibit a considerable enhancement in the on/off ratio (~10^2 times) and electron mobility (~2.5 times) compared to the MoS_2 FETs with Au/Ti electrodes. We show that the contact barrier potential of the MoS_2 FETs with graphene electrodes can be effectively tuned by applying an electric field. The work function of graphene becomes higher than that of MoS_2 at a negative bias voltage, resulting in the formation of a Schottky barrier. Similarly, the work function of graphene becomes lower than that of MoS_2 at a positive bias voltage, resulting in the formation of an ohmic barrier. The contact barrier between MoS_2 and graphene can be easily tuned using graphene electrodes. Thus, the on/off ratio and electron mobility of the MoS_2 FET can be improved by tuning the contact barrier.

2. Materials and Methods

2.1. Graphene Growth and Transfer

Graphene was synthesized on a copper foil (99.8% purity, 0.025 mm thick, Alfa Aesar, Haverhill, MA, USA) using CVD at a growth temperature of 1050 °C with 10 sccm of H_2 and 15 sccm of CH_4 [37]. Then, the full side of the foil that faced upwards during synthesis was covered with poly(methyl methacrylate) (PMMA) (AR–N 7500.18, Allresist, Strausberg, Germany) via spin coating (4000 rpm for 60 s). The remaining graphene on the Cu foil that faced downwards during the synthesis was removed using O_2 plasma (Femto, Diener, Ebhausen, Germany). The Cu foil was completely etched using 0.1 M ammonia persulfate (Sigma Aldrich, St. Louis, MO, USA). The PMMA/graphene layer was washed several times with fresh deionized water. Finally, the PMMA/graphene layer floated on the surface of the water, and it was transferred to a SiO_2 substrate. The transferred PMMA/graphene layer was patterned using electron beam lithography (Nanobeam nB4, NBL, Cambridge, UK) as shown in Figure S1a.

2.2. Fabrication of the MoS_2 Field-Effect Transistor

As shown in Figure S1b, single–layer MoS_2 was prepared via mechanical exfoliation from a bulk MoS_2 flake (429ML–AB, SPI Supplies, West Chester, PA, USA). To fabricate MoS_2 FET with graphene electrode, a dry transfer process was employed [38]. Patterned graphene was transferred onto single–layer MoS_2 flake after the alignment position using a micromanipulator (NMO–203, Narishige, Tokyo, Japan) (Figure 1a). Au/Ti electrodes were patterned using electron beam lithography with a positive electron beam resist (AR–P 671.04, Allresist, Strausberg, Germany). This was followed by metal deposition (Ti (5 nm)/Au (45 nm)) and a lift-off process.

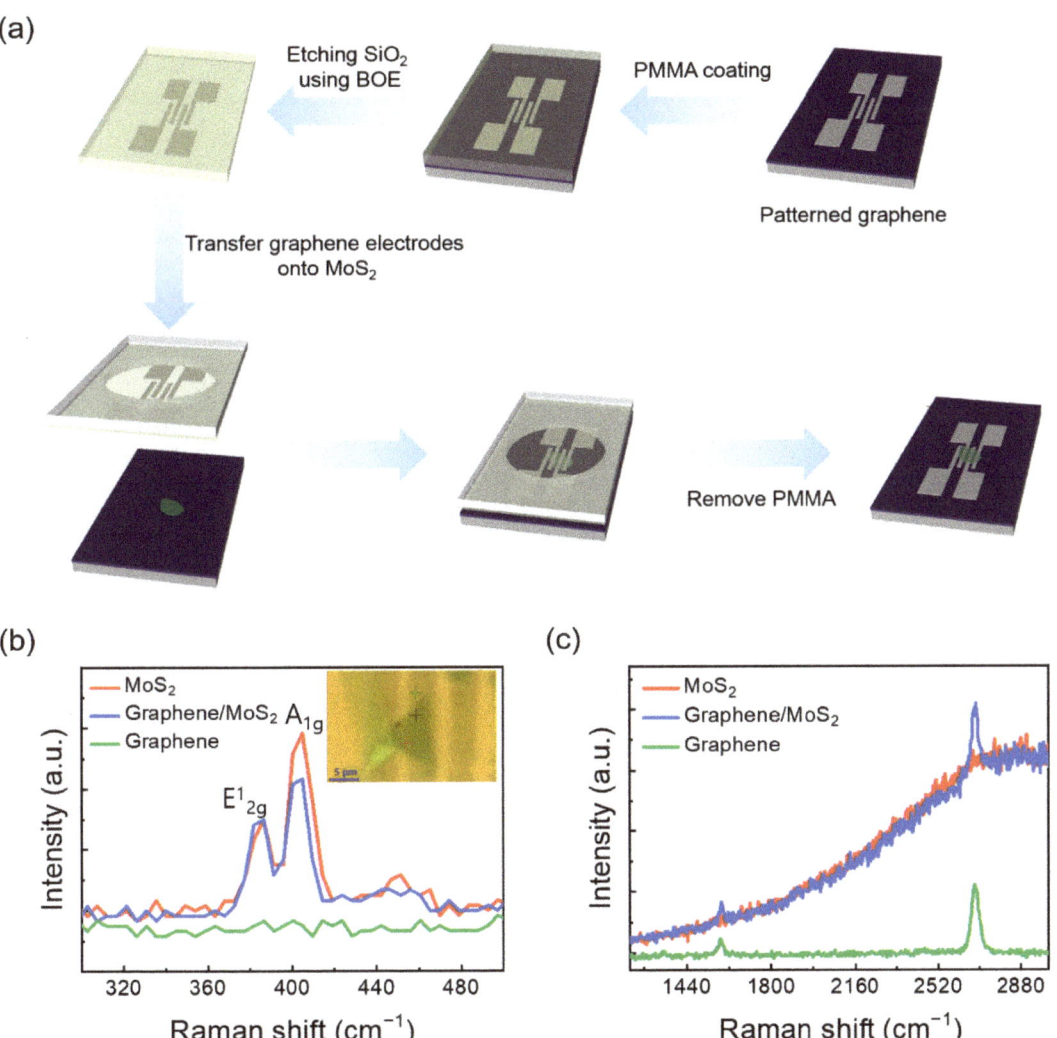

Figure 1. (a) Schematics illustration of the fabrication process for MoS_2 FET with graphene electrode. Raman spectroscopy of mechanically exfoliated single–layer MoS_2 (red), chemical vapor deposition (CVD)-grown graphene on single-layer MoS_2 (blue), and CVD-grown graphene (green). (b) MoS_2 region, and (c) graphene region of the Raman spectrum (the insert of (b) shows the Raman analysis position by cross mark).

2.3. Characterization of the MoS_2 Thin Film and Field–Effect Transistor

Mechanical exfoliation was employed to extract high-quality single-layer MoS_2 from bulk MoS_2 [3]. Then, single–layer MoS_2 was transferred onto a silicon wafer with a 300 nm thick SiO_2 layer. Raman spectroscopy (Alpha 300R, WiTec, Ulm, Germany) was used to determine the number of layers of MoS_2 [39]. The Raman spectrum of MoS_2 revealed a peak spacing of less than 20 cm^{-1} between the E_{2g} and A_{1g} modes, indicating that single-layer MoS_2 was formed. A 532 nm laser with a power of 1 mW was used as an excitation source. The exposure time was 1 s, and calibration was performed using a reference Si peak position of 520 cm^{-1}. The fabricated MoS_2 FETs were loaded into a vacuum chamber

(Lake Shore) for electrical measurements. The electrical properties of the MoS$_2$ FETs with graphene–graphene electrodes, Au/Ti–Au/Ti electrodes, and graphene–Au/Ti electrodes were characterized in vacuum (under 10^{-4} Torr) using a semiconductor parameter analyzer (4200-SCS with a preamplifier unit, Keithley, Cleveland, OH, USA) for comparison.

3. Results and Discussion

Figure 1b,c show the Raman spectra of single-layer MoS$_2$, graphene on single-layer MoS$_2$, and graphene, respectively. The MoS$_2$ and graphene/MoS$_2$ layers exhibit typical single-layer Raman active modes (~18.27 cm^{-1} of frequency difference between E$_{2g}$ and A$_{1g}$), and the 2D/G ratio of graphene is about 4.06. Therefore, it can be noted that exfoliated MoS$_2$ flake has a formation of single layer. The Dirac point of intrinsic graphene is at zero gate voltage, the work function of which is approximately 4.5 eV [32]. As shown in Figure S2, the Dirac point of the CVD-grown graphene electrode was measured at 22.5 V, owing to the hole doping originated from both coupling with dielectric layer of SiO$_2$ and exposure to oxygen and moisture [40]. The schematic of the band structure of graphene and MoS$_2$ is shown in Figure 2. Graphene and single-layer MoS$_2$ were bonded via weak vdW forces. However, MoS$_2$ and the metal interfaces formed covalent interactions, causing a change in the electronic structure [30]. This led to unexpected contact resistance. Three different types of single-layer MoS$_2$ FETs were fabricated to investigate the effects of the graphene electrode. The first was a single-layer MoS$_2$ FET with a Au/Ti–graphene heteroelectrode, as shown in Figure 3a. Highly boron-doped Si (resistance of 0.001 Ω) with a 300 nm thick SiO$_2$ layer was used as the substrate. The channel length and width of the mechanically exfoliated MoS$_2$ used in the single-layer MoS$_2$ FET were ~2 µm and ~4 µm, respectively. Figure 3b shows the asymmetric I$_{DS}$–V$_{DS}$ output characteristics of the single-layer MoS$_2$ FET with the Au/Ti–graphene heteroelectrode without the gate voltage. Different contact barriers were generated according to the contact material. An ohmic contact was formed between single-layer MoS$_2$ and Au/Ti. A Schottky contact was formed between single-layer MoS$_2$ and graphene. Figure 3c shows the I$_{DS}$–V$_g$ transfer characteristics for a positive source–drain voltage (V$_{DS}$). The on/off ratio and electron mobility (graphene in the heteroelectrode) were >10^5 and ~3.2 cm^2/V·s, respectively. Figure 3d shows the I$_{DS}$–V$_g$ transfer characteristics for a negative drain voltage. The on/off ratio and electron mobility (Au/Ti in the heteroelectrode) were >10^2 and ~1.2 cm^2/V·s, respectively. These results indicated that graphene could be used as an ideal electrode in a single-layer MoS$_2$ FET. Mobility was calculated using the following equation: $\mu_e = g_m \times L / C_g \times V_D \times W$; where g_m is the transconductance, V_D is the source–drain voltage, L is the channel length, W is the channel width, and C_g is the capacitance of 300 nm thick SiO$_2$. The MoS$_2$ FET with the Au/Ti electrodes exhibited ohmic contact behavior, whereas the MoS$_2$ FET with the graphene electrodes exhibited Schottky contact behavior. Multilayer MoS$_2$ FETs with exfoliated graphene electrodes also showed ohmic contact behavior [41]. The work function of graphene was approximately 4.5 eV because mechanically exfoliated graphene was almost pure with no doping. Therefore, the single-layer MoS$_2$ FET with the graphene electrodes exhibited a Schottky barrier without a gate bias voltage. However, the work function of graphene was electrostatically adjusted to approximately 300 meV for single-layer graphene by tuning the Fermi level (E$_F$) by changing the gate voltage by 50 V [32]. The work function of graphene decreased at a positive gate bias voltage. Figure 4 shows the I$_{DS}$–V$_{DS}$ characteristics of the single-layer MoS$_2$ FET as a function of the back-gate voltage. The Schottky barrier between graphene and single-layer MoS$_2$ was enhanced at a negative gate voltage; thus, current could not flow in the negative gate voltage direction (Figure 4a). As the gate was positively biased, the Schottky barrier between graphene and single-layer MoS$_2$ decreased, and the contact barrier between single-layer MoS$_2$ and Au/Ti did not change. The I$_{DS}$–V$_{DS}$ output characteristics of the single-layer MoS$_2$ FET with the Au/Ti–graphene heteroelectrode (green solid line) showed almost similar with linear (red dashed line) at a gate voltage of 20 V because the work function of graphene became similar to that of single-layer MoS$_2$ (Figure 4c). As the gate voltage exceeded 20 V, the current level (black solid line) of the graphene electrode became higher than that of the Au/Ti electrode (Figure 4d). These results showed that the electrical properties of the single-layer MoS$_2$ FET were enhanced using the graphene electrodes. A Schottky barrier was formed at the interface of graphene and MoS$_2$ in the current-off region; thus, there was no leakage current. However, an ohmic barrier was formed at the interface between graphene and MoS$_2$ in the current-on region. Therefore, the on/off ratio and electron mobility of single-layer MoS$_2$ were high. The on/off ratio and electron mobility of single–layer MoS$_2$ were compared with those of homogeneous electrodes. A single-layer MoS$_2$ FET with the graphene electrodes was fabricated, and its electrical properties were measured. Figure 5a shows the schematic of the single-layer MoS$_2$ FET with the graphene electrodes, and Figure 5b shows its I_{DS}–V_g transfer characteristics. The I_{DS}–V_{DS} output characteristics shown in Figure 5c confirmed that a Schottky barrier was formed. When an increasingly positive back-gate bias was applied to the single-layer MoS$_2$ FET with the graphene electrodes, the Schottky barrier was slightly modified into a clear ohmic contact, as shown in Figure 5d. The on/off ratio and electron mobility were >10^5 and ~2.3 cm^2/V·s, respectively. A single-layer MoS$_2$ FET with the Au/Ti electrodes was fabricated, and its electrical properties were measured for comparison. Figure S2a shows the schematic of the single-layer MoS$_2$ FET with the Au/Ti electrodes, and Figure S2b shows its I_{DS}–V_g transfer characteristics. The on/off ratio and electron mobility were >10^3 and ~0.9 cm^2/V·s, respectively. The on/off ratio and electron mobility of the single-layer MoS$_2$ FET with the graphene electrodes were ~10^2 and ~2.5 times higher than those of the single-layer MoS$_2$ FET with the Au/Ti electrodes, respectively. To study the barrier height of the MoS$_2$ FET with graphene electrode, current voltage characteristics (Figure 6a) and I_{DS}–V_g transfer characteristics (Figure 6b) were measured at different temperatures. The 2D thermionic emission equation was used to describe the electrical transport behavior of Schottky contacted MoS$_2$ devices [41,42].

$$I_{DS} = AA^*_{2D}T^{3/2}\exp\left[\frac{q}{k_B T}\left(\Phi_B - \frac{V_{DS}}{n}\right)\right] \quad (1)$$

where A is the contact area of the junction, A^*_{2D} is the two–dimensional equivalent Richardson constant, q is the magnitude of the electron charge, Φ_B is the Schottky barrier height, k_B is the Boltzmann constant, n is the ideality factor, and V_{DS} is the drain-source bias. Instead of the typical Arrehenius plot, $\ln(I_d/T^2)$ versus $1000/T$ for three-dimensional semiconductors, $\ln(I_d/T^{3/2})$ versus $1000/T$ was used because here the semiconducting channel is two-dimensional. The $\ln(I_d/T^{3/2})$ versus $1000/T$ of MoS$_2$ FET with graphene electrodes for various values of V_g is shown in Figure 6c. Based on Equation (1), the height of the Schottky barrier can be deduced as Equation (2):

$$y_{intercept} = -\frac{q}{1000k_B}\Phi_B \quad (2)$$

In the MoS$_2$ FET with graphene electrodes, the Schottky barrier is decreased dramatically—from 51.5 meV to 0 meV—with the back gate voltage changing from −7.5 to 12.5 V, as shown in Figure 6d. The change of the Schottky barrier in the MoS$_2$ FET with graphene electrodes comes from changes in work function of graphene.

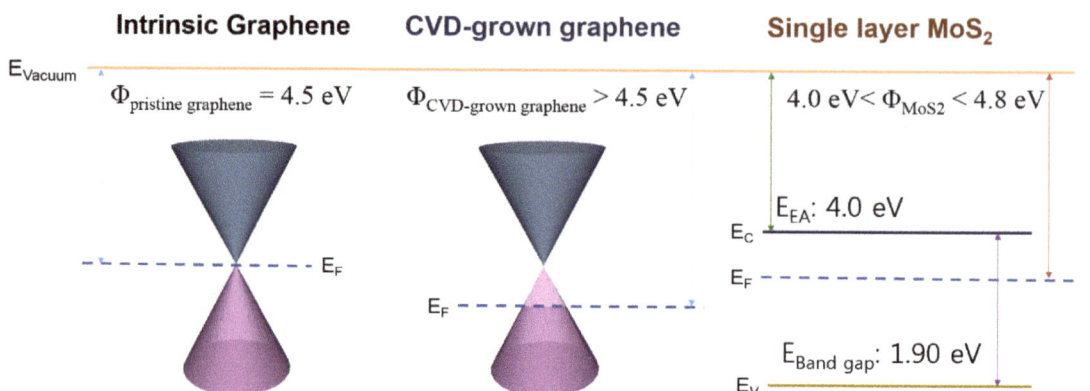

Figure 2. Schematic band diagram of intrinsic graphene, CVD-grown graphene, and single-layer MoS$_2$.

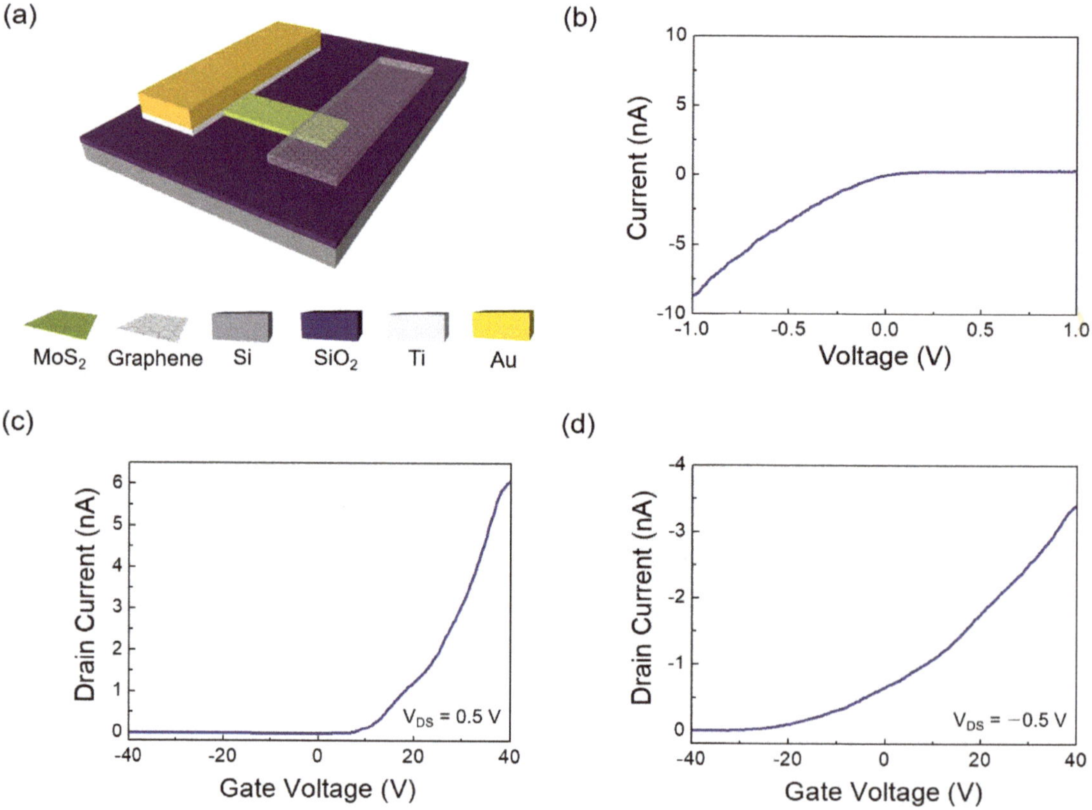

Figure 3. Schematic and electrical properties of MoS$_2$ field-effect transistor (FET) with hetero-electrodes. (**a**) Schematic of MoS$_2$ FET with heteroelectrodes; (**b**) I$_{DS}$–V$_{DS}$ output characteristics; (**c**) I$_{DS}$–V$_g$ transfer characteristics at V$_{DS}$ = 0.5 V; (**d**) I$_{DS}$–V$_g$ transfer characteristics at V$_{DS}$ = −0.5 V.

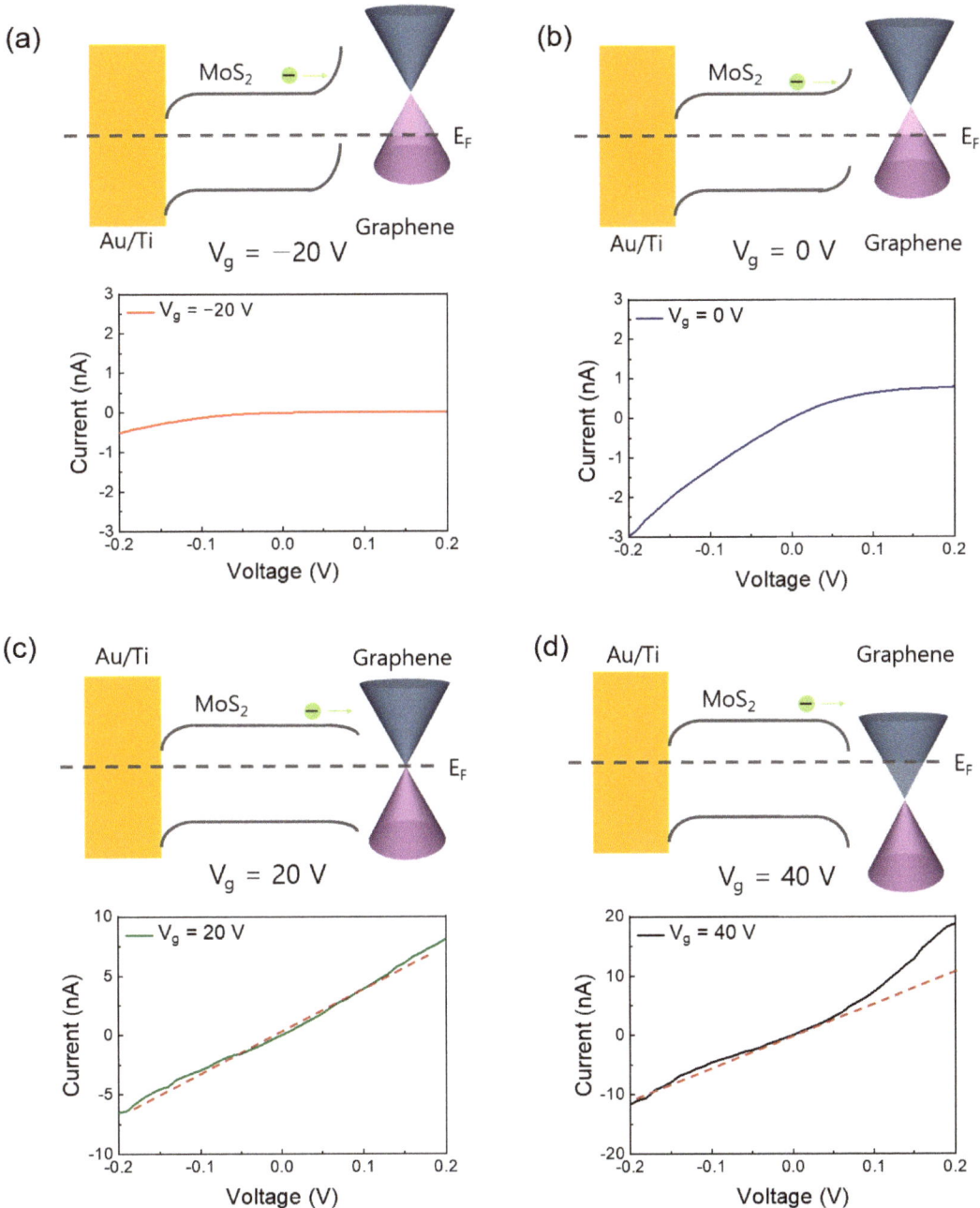

Figure 4. Band diagrams and electrical properties of the MoS$_2$ FET with Au–graphene heteroelectrode at different gate voltages ((**a**) −20 V, (**b**) 0 V, (**c**) 20 V, and (**d**) 40 V).

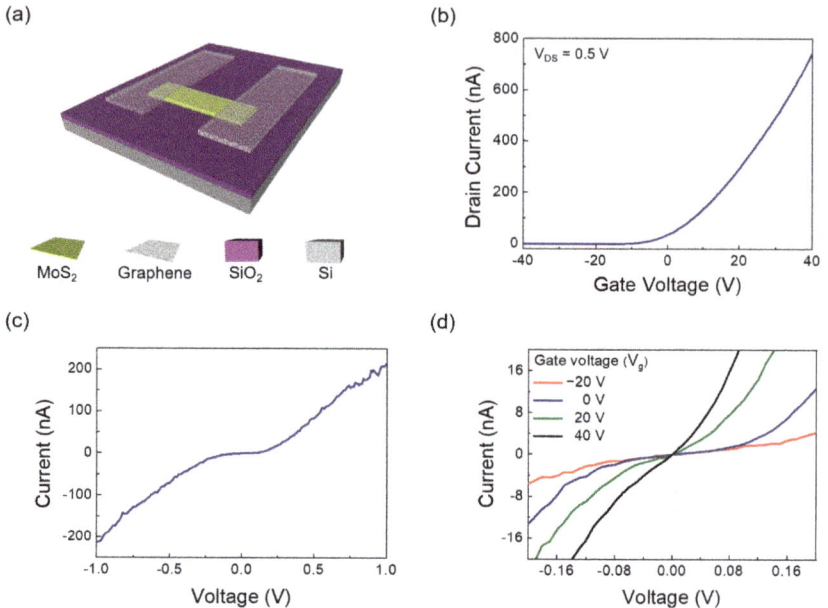

Figure 5. Schematic and electrical properties of MoS$_2$ FET with graphene electrodes. (**a**) Schematic of MoS$_2$ FET with graphene electrodes; (**b**) I$_{DS}$–V$_g$ transfer characteristics at V$_{DS}$ = 0.5 V; (**c**) I$_{DS}$–V$_{DS}$ output characteristics; (**d**) I$_{DS}$–V$_{DS}$ characteristics at different gate bias voltages.

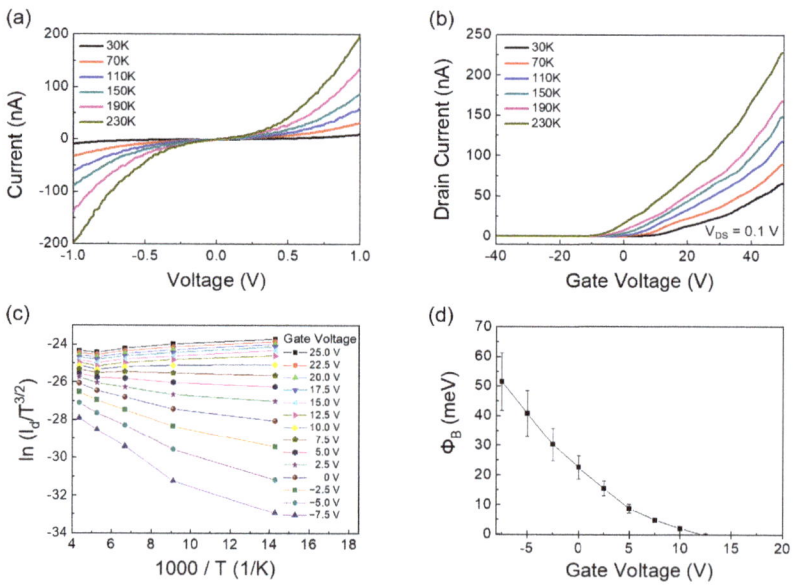

Figure 6. Temperature-dependent electrical transport of the MoS$_2$ FET with graphene electrode. (**a**) Current voltage characteristics and (**b**) I$_{DS}$–V$_g$ transfer characteristics from 30 K to 230 K, for source-drain bias voltage of 0.1 V. (**c**) Linear fit of the Arrhenius plot, $\ln\left(I_d/T^{3/2}\right)$ vs. $1000/T$ as function of V$_g$. (**d**) The Schottky barrier of MoS$_2$ FET with graphene electrode depends on the gate voltage.

4. Conclusions

This work demonstrates the enhancement of the electrical properties of an MoS$_2$ FET with graphene electrodes by tuning the contact barrier using an electric field. The MoS$_2$ FET with a Au/Ti–graphene heteroelectrode shows a clear change in the contact barrier between MoS$_2$ and graphene. A Schottky barrier and ohmic barrier exist in the off and on states of the MoS$_2$ FET with the graphene electrodes. The on/off ratio and electron mobility of the MoS$_2$ FET with the graphene electrodes are 10^2 and 2.5 times higher than those of the MoS$_2$ FET with the Au/Ti electrodes, respectively. The Schottky barrier between MoS$_2$ and graphene is decreased from 51.5 to 0 meV by the back gate voltage. The implication of these results could be of great importance in better understanding the desirable contact formation for high performance MoS$_2$ FETs. This FET may be promising for electronic device applications based on next-generation 2D materials.

Supplementary Materials: The following supporting information can be downloaded at: https://www.mdpi.com/article/10.3390/nano12173038/s1, Figure S1: Schematic illustration of sample preparation process; Figure S2: Electrical properties of chemical-vapor-deposition-grown graphene; Figure S3: Schematic and electrical properties of MoS$_2$ field-effect transistor with Au/Ti electrodes.

Funding: This research was supported by the National Research Foundation of Korea (NRF) grant funded by the Korea government (MSIT) (2021R1A4A1031900, 2021R1I1A3049729) and Nano·Material Technology Development Program through the National Research Foundation of Korea (NRF) funded by the Ministry of Science, ICT and Future Planning (N220212001).

Institutional Review Board Statement: Not applicable.

Informed Consent Statement: Not applicable.

Data Availability Statement: Not applicable.

Conflicts of Interest: The author declare no conflict of interest.

References

1. Glavin, N.R.; Rao, R.; Varshney, V.; Bianco, E.; Apte, A.; Roy, A.; Ringe, E.; Ajayan, P.M. 2D materials: Emerging applications of elemental 2D materials. *Adv. Mater.* **2020**, *32*, 1904302. [CrossRef] [PubMed]
2. Cheng, Z.; Cao, R.; Wei, K.; Yao, Y.; Liu, X.; Kang, J.; Dong, J.; Shi, Z.; Zhang, H.; Zhang, X. 2D materials enabled next-generation integrated optoelectronics: From fabrication to applications. *Adv. Sci.* **2021**, *8*, 2003834. [CrossRef]
3. Novoselov, K.S.; Geim, A.K.; Morozov, S.V.; Jiang, D.; Zhang, Y.; Dubonos, S.V.; Grigorieva, I.V.; Firsov, A.A. Electric field effect in atomically thin carbon films. *Science* **2004**, *306*, 666–669. [CrossRef] [PubMed]
4. Novoselov, K.S.; Geim, A.K.; Morozov, S.V.; Jiang, D.; Katsnelson, M.I.; Grigorieva, I.V.; Dubonos, S.V.; Firsov, A.A. Two-dimensional gas of massless Dirac fermions in graphene. *Nature* **2005**, *438*, 197–200. [CrossRef] [PubMed]
5. Zhang, Y.; Tan, Y.-W.; Srormer, H.L.; Kim, P. Experimental observation of the quantum Hall effect and Berry's phase in graphene. *Nature* **2005**, *438*, 201–204. [CrossRef]
6. Meric, I.; Han, M.Y.; Young, A.F.; Ozylmaz, B.; Kim, P.; Shepard, K.L. Current saturation in zero-bandgap, top-gated graphene field-effect transistors. *Nat. Nanotechnol.* **2008**, *3*, 654–659. [CrossRef]
7. Li, X.; Wang, X.; Zhang, L.; Lee, S.; Dai, H. Chemically Derived. Ultrasmooth graphene nanoribbon semiconductors. *Science* **2008**, *319*, 1229–1232. [CrossRef]
8. Wang, X.; Ouyang, Y.; Li, X.; Wang, H.; Guo, J.; Dai, H. Room-temperature all-semiconducting sub-10-nm graphene nanoribbon field-effect-transistors. *Phys. Rev. Lett.* **2008**, *100*, 206803. [CrossRef]
9. Jing, L.; Velasco, J., Jr.; Kratz, P.; Liu, G.; Bao, W.; Bockrath, M.; Lau, C.N. Quantum transport and field-induced insulating states in bilayer graphene pnp junctions. *Nano Lett.* **2010**, *10*, 4000–4004. [CrossRef]
10. Castro, E.V.; Novoselov, K.S.; Morozov, S.V.; Peres, N.M.R.; Santos, J.M.B.L.; Nilsson, J.; Guinea, F.; Geim, A.K.; Neto, A.H.C. Biased bilayer graphene: Semiconductor with a gap tunable by the electric field effect. *Phys. Rev. Lett.* **2007**, *99*, 216802. [CrossRef]
11. Taychatanapat, T.; Jarillo-Herrero, P. Electronic transport in Dual-gated bilayer graphene at large displacement fields. *Phys. Rev. Lett.* **2010**, *105*, 166601. [CrossRef] [PubMed]
12. Xia, F.; Farmer, D.B.; Lin, Y.-M.; Avouris, P. Graphene field-effect transistors with high On/Off current ratio and large transport band gap at room temperature. *Nano Lett.* **2010**, *10*, 715–718. [CrossRef] [PubMed]
13. Yang, H.; Heo, J.; Park, S.; Song, H.J.; Seo, D.H.; Byun, K.-E.; Kim, P.; Yoo, I.; Chung, H.-J.; Kim, K. Graphene barrister, a triode device with a gate-controlled Schottky barrier. *Science* **2012**, *336*, 1140–1143. [CrossRef] [PubMed]

14. Britnell, L.; Gorbachev, R.V.; Jalil, R.; Bello, B.D.; Schedin, F.; Mishchenko, A.; Georgiou, T.; Katsnelson, M.I.; Eaves, L.; Morozov, S.V.; et al. Field-Effect Tunneling Transistor Based on Vertical Graphene Heterostructures. *Science* **2012**, *335*, 947–950. [CrossRef] [PubMed]
15. Mak, K.F.; Lee, C.; Hone, J.; Shan, J.; Heinz, T.F. Atomically thin MoS_2: A new direct-gap semiconductor. *Phys. Rev. Lett.* **2010**, *105*, 136805. [CrossRef]
16. Eda, G.; Yamaguchi, H.; Voiry, D.; Fujita, T.; Chen, M.; Chhowalla, M. Photoluminescence from Chemically Exfoliated MoS_2. *Nano Lett.* **2011**, *11*, 5111–5116. [CrossRef]
17. Lee, H.S.; Min, S.-W.; Chang, Y.-G.; Park, M.K.; Nam, T.; Kim, H.; Kim, J.H.; Ryu, S.; Im, S. MoS_2 Nanosheet Phototransistors with Thickness-Modulated Optical Energy Gap. *Nano Lett.* **2012**, *12*, 3695–3700. [CrossRef]
18. Najmaei, S.; Liu, Z.; Zhou, W.; Zou, X.; Shi, G.; Lei, S.; Yakobson, B.I.; Idrobo, J.-C.; Ajayan, P.M.; Lou, J. Vapour phase growth and grain boundary structure of molybdenum disulphide atomic layers. *Nat. Mater.* **2013**, *12*, 754–759. [CrossRef]
19. Lee, Y.-H.; Zhang, X.-Q.; Zhang, W.; Chang, M.-T.; Lin, C.-T.; Chang, K.-D.; Yu, Y.-C.; Wang, J.T.-W.; Chang, C.-S.; Li, L.-J.; et al. Synthesis of large-area MoS_2 atomic layers with chemical vapor deposition. *Adv. Mater.* **2012**, *24*, 2320–2325. [CrossRef]
20. Lee, Y.-H.; Yu, L.; Wang, H.; Fang, W.; Ling, X.; Shi, Y.; Lin, C.-T.; Huang, J.-K.; Chang, M.-T.; Chang, C.-S.; et al. Synthesis and Transfer of Single-Layer Transition Metal Disulfides on Diverse Surfaces. *Nano Lett.* **2013**, *13*, 1852–1857. [CrossRef]
21. Ling, X.; Lee, Y.-H.; Lin, Y.; Fang, W.; Yu, L.; Dresselhaus, M.S.; Kong, J. Role of the Seeding Promoter in MoS_2 Growth by Chemical Vapor Deposition. *Nano Lett.* **2014**, *14*, 464–472. [CrossRef] [PubMed]
22. Schmidit, H.; Wang, S.; Chu, L.; Toh, M.; Kumar, R.; Zhao, W.; Neto, A.H.C.; Martin, J.; Adam, S.; Adam, S.; et al. Transport Properties of Monolayer MoS2 Grown by Chemical Vapor Deposition. *Nano Lett.* **2014**, *14*, 1909–1913. [CrossRef]
23. Schulman, D.S.; Arnold, A.J.; Das, S. Contact engineering for 2D materials and devices. *Chem. Soc. Rev.* **2018**, *47*, 3037–3058. [CrossRef] [PubMed]
24. Chai, J.W.; Yang, M.; Callsen, M.; Zhou, J.; Yang, T.; Zhang, Z.; Pan, J.S.; Chi, D.Z.; Feng, Y.P.; Wang, S.J. Tuning contact barrier height between metals and MoS2 monolayer through interface engineering. *Adv. Mater. Interfaces* **2017**, *4*, 170035. [CrossRef]
25. Kaasbjerg, K.; Thygesen, K.S.; Jacobsen, K.W. Phonon-limited mobility in n-type single-layer MoS_2 from first principles. *Phys. Rev. B* **2012**, *85*, 115317. [CrossRef]
26. Radisavljevic, B.; Kis, A. Mobility engineering and a metal-insulator transition in monolayer MoS_2. *Nat. Mater.* **2013**, *12*, 815–820. [CrossRef]
27. Jariwala, D.; Sangwan, V.K.; Late, D.; Johns, J.E.; Dravid, V.P.; Marks, T.J.; Lauhon, L.J.; Hersam, M.C. Band-like transport in high mobility unencapsulated single-layer MoS_2 transistors. *Appl. Phys. Lett.* **2013**, *102*, 173107. [CrossRef]
28. Qiu, H.; Xu, T.; Wang, Z.; Ren, W.; Nan, H.; Ni, Z.; Chen, Q.; Yuan, S.; Miao, F.; Song, F.; et al. Hopping transport through defect-induced localized states in molybdenum disulphide. *Nat. Commun.* **2013**, *4*, 2642. [CrossRef]
29. Popov, I.; Seifert, G.; Tomanek, D. Designing Electrical Contacts to MoS_2 Monolayers: A Computational Study. *Phys. Rev. Lett.* **2012**, *108*, 156802. [CrossRef]
30. Gong, C.; Colombo, L.; Wallace, R.M.; Cho, K. The Unusual Mechanism of Partial Fermi Level Pinning at Metal-MoS_2 Interfaces. *Nano Lett.* **2014**, *14*, 1714–1720. [CrossRef]
31. Gong, C.; Zhang, H.; Wang, W.; Colombo, L.; Wallace, R.M.; Cho, K. Band alignment of two-dimensional transition metal dichalcogenides: Application in tunnel field effect transistors. *Appl. Phys. Lett.* **2013**, *103*, 053513. [CrossRef]
32. Yu, Y.-J.; Zhao, Y.; Ryu, S.; Brus, L.E.; Kim, K.S.; Kim, P. Tuning the Graphene Work Function by Electric Field Effect. *Nano Lett.* **2009**, *9*, 3430–3434. [CrossRef] [PubMed]
33. Huang, H.; Xu, W.; Chen, T.; Chang, R.-J.; Sheng, Y.; Zhang, Q.; Hou, L.; Warner, J.H. High-Performance Two-Dimensional Schottky Diodes Utilizing Chemical Vapour Deposition-Grown Graphene-MoS_2 Heterojunctions. *ACS Appl. Mater. Interfaces* **2018**, *10*, 37258–37266. [CrossRef]
34. Baik, S.S.; Im, S.; Choi, H.J. Work Function Tuning in Two-Dimensional MoS_2 Field-Effect-Transistors with Graphene and Titanium Source-Drain Contacts. *Sci. Rep.* **2017**, *7*, 45546. [CrossRef] [PubMed]
35. Tian, H.; Tan, Z.; Wu, C.; Wang, X.; Mohammad, M.A.; Xie, D.; Yang, Y.; Wang, J.; Li, L.-J.; Xu, J.; et al. Novel Field-Effect Schottky Barrier Transistors Based on Graphene-MoS_2 Heterojunctions. *Sci. Rep.* **2014**, *4*, 5951. [CrossRef] [PubMed]
36. Shin, C.-J.; Wang, Q.H.; Son, Y.; Jin, Z.; Blankschtein, D.; Strano, M.S. Tuning On-Off Current Ratio and Field-Effect Mobility in a MoS_2-Graphene Heterostructure via Schottky Barrier Modulation. *ACS Nano* **2014**, *8*, 5790–5798.
37. Li, X.; Cai, W.; An, J.; Kim, S.; Nah, J.; Yang, D.; Piner, R.; Velamakanni, A.; Jung, I.; Tutuc, E.; et al. Large-Area Synthesis of High-Quality and Uniform Graphene Films on Copper Foils. *Science* **2009**, *324*, 1312–1314. [CrossRef]
38. Dean, C.R.; Young, A.F.; Meric, I.; Lee, C.; Wang, L.; Sorgenfrei, S.; Watanabe, K.; Taniguchi, T.; Kim, P.; Shepard, K.L.; et al. Boron nitride substrates for high-quality graphene electronics. *Nat. Nanotechnol.* **2010**, *5*, 722–726. [CrossRef]
39. Li, H.; Zhang, Q.; Yap, C.C.R.; Tay, B.K.; Edwin, T.H.T.; Olivier, A.; Baillargeat, D. From Bulk to Monolayer MoS_2: Evolution of Raman Scattering. *Adv. Funct. Mater.* **2012**, *22*, 1385–1390. [CrossRef]
40. Ryu, S.; Liu, L.; Berciaud, S.; Yu, Y.-J.; Liu, H.; Kim, P.; Flynn, G.W.; Brus, L.E. Atmospheric Oxygen Binding and Hole Doping in Deformed Graphene on a SiO_2 Substarte. *Nano Lett.* **2010**, *10*, 4944–4951. [CrossRef]

41. Yoon, J.; Park, W.; Bae, G.-Y.; Kim, Y.; Jang, H.S.; Hyun, Y.; Lim, S.K.; Kahng, Y.H.; Hong, W.-K.; Lee, B.H.; et al. Highly Flexible and Transparent Multilayer MoS_2 Transistors with Graphene Electrodes. *Small* **2013**, *9*, 3295–3300. [CrossRef] [PubMed]
42. Chen, J.-R.; Odenthal, P.M.; Swartz, A.G.; Floyd, G.C.; Wen, H.; Luo, K.Y.; Kawakami, R.K. Control of Schottky Barriers in Single Layer MoS2 Transistors with Ferromagnetic Contacts. *Nano Lett.* **2013**, *13*, 3106–3110. [PubMed]

Article

Controlled p-Type Doping of MoS$_2$ Monolayer by Copper Chloride

Sangyeon Pak

School of Electronic and Electrical Engineering, Hongik University, Seoul 04066, Korea; spak@hongik.ac.kr

Abstract: Electronic devices based on two-dimensional (2D) MoS$_2$ show great promise as future building blocks in electronic circuits due to their outstanding electrical, optical, and mechanical properties. Despite the high importance of doping of these 2D materials for designing field-effect transistors (FETs) and logic circuits, a simple and controllable doping methodology still needs to be developed in order to tailor their device properties. Here, we found a simple and effective chemical doping strategy for MoS$_2$ monolayers using CuCl$_2$ solution. The CuCl$_2$ solution was simply spin-coated on MoS$_2$ with different concentrations under ambient conditions for effectively p-doping the MoS$_2$ monolayers. This was systematically analyzed using various spectroscopic measurements using Raman, photoluminescence, and X-ray photoelectron and electrical measurements by observing the change in transfer and output characteristics of MoS$_2$ FETs before and after CuCl$_2$ doping, showing effective p-type doping behaviors as observed through the shift of threshold voltages (Vth) and reducing the ON and OFF current level. Our results open the possibility of providing effective and simple doping strategies for 2D materials and other nanomaterials without causing any detrimental damage.

Keywords: MoS$_2$ monolayer; copper chloride; transition metal chloride; p-type doping; spin coating

1. Introduction

Monolayer transition metal dichalcogenides (TMDCs) have been considered to be the next-generation semiconducting channel materials because of their incredible electronic and mechanical properties that make them suitable for flexible, wearable, and transparent devices [1–5]. In addition, their ideally dangling bond-free surface and atomic thickness show promise for van der Waals integration on various substrates/materials and in reducing short channel effect, thus becoming candidates for the semiconducting channel materials in nano-scaled electronics and optoelectronics devices [2,6–8]. Especially, the field-effect transistors (FETs) composed of TMDC monolayers show high carrier mobility, large On/Off ratio (>10^8), and low power consumption, which have inspired experimental research in advancing FET performance of these devices [5,9–14].

To implement the TMDC monolayers for practical electronic device applications, the device properties need to be tailored to show the desired output characteristics of electronic devices. One way to achieve the desired device characteristics is through doping [3,15–19]. The doping of 2D materials is recognized to be the key to precisely controlling their fundamental properties, based on the history of the contemporary Si or III–V-based semiconductors. Ion implantation is one of the possible doping techniques. However, this uses high energy and can be detrimental to atomically thin 2D crystals. On the other hand, chemical doping is potentially more advantageous compared to the ion implantation method, as the chemical doping is generally based on a charge transfer by chemical potential of adsorbed organic molecules and leads to less damage in 2D crystal structures [18,20–23]. The chemical doping of 2D TMDCs have been mostly relied on employing self-assemble monolayer (SAM) techniques [3,9], substitutional doping [17,24], and passivation of sulfur vacancy defects [5,25,26]. However, using such techniques, it is difficult to tune the amount

of doping, and they generally require a controlled doping environment. Therefore, it is required to find a convenient method to modulate electrical/optical properties, as well as the electronic device properties.

Herein, we report a simple and controllable doping method for 2D MoS_2 using copper (II) chloride ($CuCl_2$) performed at ambient conditions. In this process, the $CuCl_2$ is dissolved in ethanol, and the solution is simply spin-coated onto 2D MoS_2 to effectively modulate charge carrier densities without any damage to MoS_2 crystals and their devices. The change in doping was analytically confirmed through Raman, photoluminescence (PL), and X-ray photoelectron spectroscopy (XPS), showing the p-type doping effect on 2D MoS_2. We further confirmed the feasibility of this doping process by simply coating the different concentrations of $CuCl_2$ solution onto the back-gated MoS_2 transistors, showing effective p-type doping behaviors as observed through the shift of threshold voltages (Vth) and reducing the ON and OFF current levels. These findings pave an important pathway toward modulating 2D materials and devices and designing logic devices based on 2D materials.

2. Materials and Methods

Synthesis of monolayer MoS_2: Monolayer MoS_2 was synthesized on a SiO_2 (300 nm)/Si substrate using the previously reported thermal chemical vapor deposition (CVD) method [27,28]. Here, 0.05 mg of MoO_3 precursors were prepared by dissolving the MoO_3 powders into ammonium hydroxide (NH_4OH) solution and loading the MoO_3 onto alumina boats using a micropipette. The SiO_2/Si substrate was placed above the alumina boat with the substrate placed faced down. The growth was carried in a 2-inch quartz tube, an alumina boat containing 100 mg of sulfur powder was placed upstream, while the alumina boat containing 0.05 mg of MoO_3, and substrate was placed downstream in the middle of the CVD furnace. The growth was carried out at 750 °C for 10 min, and the furnace was naturally cooled down to room temperature. MoS_2 crystal sizes around 30–50 μm were obtained from the CVD synthesis.

Fabrication and measurement of MoS_2 transistors: The synthesized MoS_2 monolayers were transferred onto HfO_2/Si substrate using polystyrene (PS) film as the transferring medium. The PS film (M_W~192,000) was spin-coated onto MoS_2/SiO_2/Si substrate and the film was detached during the transfer process while the film was floated on DI water. The detached PS film with MoS_2 was dried in air for 2 h and transferred onto HfO_2/Si substrate for device fabrication. The source and drain electrode pads were patterned using photolithography, and 5 nm Ti/40 nm Au electrodes were deposited using a thermal evaporator. The devices were annealed at 150 °C for 1 h under vacuum conditions. The electrical properties were measured using semiconductor parameter analyzer (Keithley 4200A-SCS) and MS Tech probe station.

Characterization of MoS_2: The Raman and PL measurements were carried out using Alpha 300 R confocal Raman spectroscopy with 532 nm laser. AFM measurement was carried out using XE7 (Park Systems, Suwon, Korea). XPS measurement was performed using NEXSA (Thermofisher Scientific, Waltham, MA, USA).

3. Results

MoS_2 monolayers were synthesized on a SiO_2 (300 nm)/Si substrate using a chemical vapor deposition (CVD). Figure 1a shows the CVD-grown monolayered MoS_2 profiled by atomic force microscopy (AFM) height measurement. The morphology and thickness of the as-grown MoS_2 show its thickness around 0.7 nm, confirming the single-layered thickness. For doping of MoS_2, $CuCl_2$ was employed in this study as the metal chlorides offer a wide range of doping molecules, have been frequently employed to modulate electrical properties of graphene, and are known to be strong electron acceptors [29–31]. The metal chloride generally acts as a strong electron acceptor due to the high electronegativity of chlorine compared to molybdenum or sulfur [31]. It should be noted that $CuCl_2$ has never been used for doping 2D MoS_2. Figure 1b illustrates our simple $CuCl_2$ doping process. $CuCl_2$ was firstly dissolved in ethanol at different molar concentrations (0.5 M and 1 M).

The as-grown MoS$_2$ layer on SiO$_2$ was then placed on a spin coater, and CuCl$_2$ solution was dropped onto the as-grown MoS$_2$, followed by the spin coating at 3000 RPM. The doped MoS$_2$ samples were then dried on a hot plate at a mild temperature below 90 °C. All of the doping processes were performed in ambient conditions.

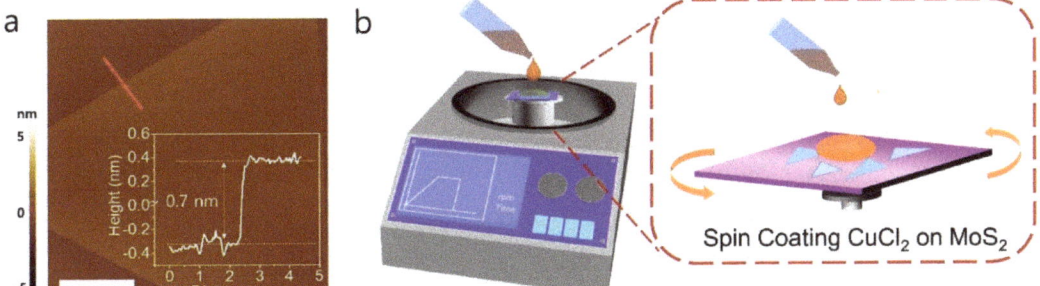

Figure 1. CuCl$_2$ doping process for MoS$_2$ monolayers. (**a**) AFM topography image of MoS$_2$ monolayers showing its atomic thickness around 0.7 nm. Scale bar: 5 µm. (**b**) Schematic illustration of CuCl$_2$ doping process. The as-synthesized MoS$_2$ monolayers on SiO$_2$/Si substrate were placed on a spin coater, and CuCl$_2$ solution was dropped onto the substrate for spin coating at 3000 RPM.

In order to understand the effect of CuCl$_2$ doping on MoS$_2$ monolayers, we first performed Raman and PL analysis of the pristine MoS$_2$ and CuCl$_2$-doped MoS$_2$ as shown in Figure 2a,b. Figure 2a shows the Raman spectrum of pristine MoS$_2$ (dotted grey line), 0.5 M CuCl$_2$-doped MoS$_2$ (purple line), and 1 M CuCl$_2$-doped MoS$_2$ (magenta). The Raman spectrum of pristine MoS$_2$ shows two characteristic peaks located at around 381 cm^{-1} and 400 cm^{-1}, which correspond to the in-plane E$^1_{2g}$ and out-of-plane A$_{1g}$ vibrational modes, respectively. As the CuCl$_2$ is doped onto MoS$_2$, the Raman peaks of MoS$_2$ were monotonically blue shifted with increasing the CuCl$_2$ doping concentrations. Such a trend of shifting in Raman peaks is a clear signature that carrier concentrations were changed without damaging the crystal structure, and can be understood as a CuCl$_2$-induced p-type doping effect, which is in agreement with the previous studies [9,32]. Figure 2b shows PL spectra measured for the pristine MoS$_2$ and CuCl$_2$-doped MoS$_2$. The pristine MoS$_2$ shows direct bandgap PL emission at around 1.82 eV. As the MoS$_2$ monolayers were doped with CuCl$_2$, the PL intensity was largely increased. It has been widely accepted that the PL intensity of 2D MoS$_2$ is strongly affected by carrier concentrations. The increased PL intensity of MoS$_2$ monolayer can be due to the reduced trion formation and strongly increased exciton radiative recombination rates through decreasing the carrier concentrations of CuCl$_2$-doped MoS$_2$ monolayers [33]. Therefore, CuCl$_2$ doping of MoS$_2$ monolayer decreases carrier concentrations due to the strong electron accepting nature of CuCl$_2$, which was observed through Raman and PL measurements.

To further confirm the effect of CuCl$_2$ doping and the chemical state of CuCl$_2$ molecules, we performed X-ray photoelectron (XPS) analysis as shown in Figure 3a–d. XPS analysis was performed for both pristine MoS$_2$ and CuCl$_2$-doped MoS$_2$, and we compared any change in the binding energies. Figure 3a,b show the main binding energy of MoS$_2$, Mo 3d and S 2p peaks, and we compared the change of binding energies before and after CuCl$_2$ doping. It is clearly observable that the binding energies of both Mo 3d and S 2p peaks shifted toward lower binding energy by about 0.3 eV. The shift to a lower binding energy in semiconducting MoS$_2$ can be attributed to the shift of Fermi-level energy toward the valence band, which results in the change of binding energies in MoS$_2$, and the results agree with the Raman and PL analysis that show the p-type doping effect of CuCl$_2$ on MoS$_2$.

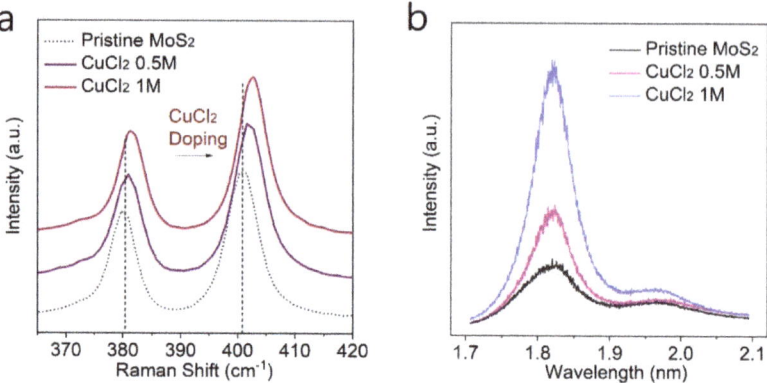

Figure 2. (**a**) Raman and (**b**) PL measurements before and after CuCl$_2$ doping. 0.5 M and 1 M CuCl$_2$ solutions were employed for doping. After CuCl$_2$ doping, the Raman spectrum of MoS$_2$ was blue-shifted and PL intensity was largely increased, showing p-type doping effect on MoS$_2$.

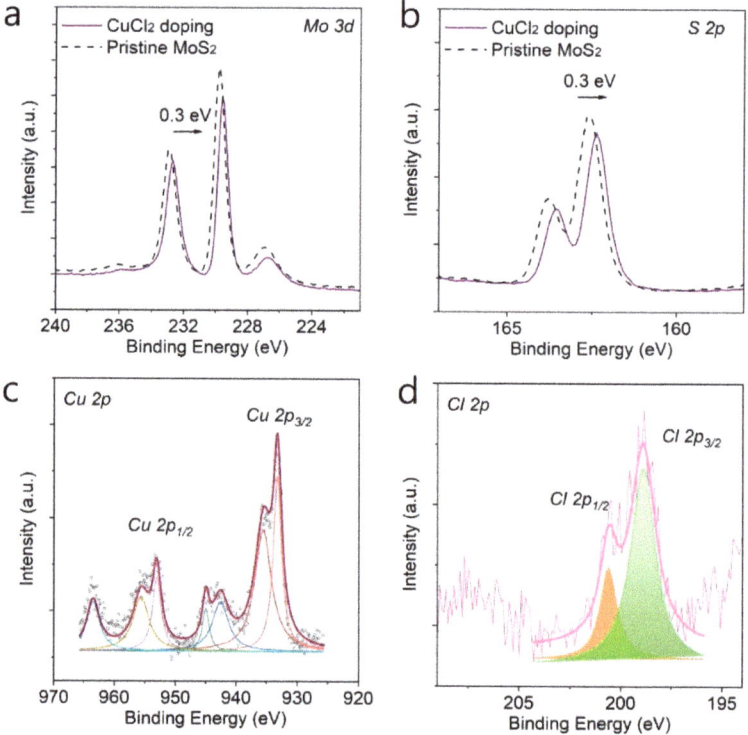

Figure 3. X-ray photoelectron spectroscopy (XPS) measurements before and after CuCl$_2$ doping. XPS spectrum of (**a**) Mo 3d and (**b**) S 2p before and after CuCl$_2$ doping. The shift of the binding energies to lower energy indicates lowered Fermi level in MoS$_2$. XPS spectrum of (**c**) Cu 2p and (**d**) Cl 2p was found in CuCl$_2$ doped MoS$_2$ film, demonstrating the CuCl$_2$ is doped onto MoS$_2$.

XPS analysis on CuCl$_2$-doped MoS$_2$ also showed the chemical state of CuCl$_2$ as shown in Figure 3c,d. Figure 3c and d show the high-resolution XPS peaks of Cu 2p and Cl 2p peaks, respectively, which were recorded from the CuCl$_2$-doped MoS$_2$ sample. As shown in Figure 3c, it can be seen that the binding energies of Cu 2p are composed of the main characteristic doublet peaks centered at around 953 eV and 933.3 eV, which correspond to Cu $2p_{1/2}$ and Cu $2p_{3/2}$, respectively, and other satellite peaks [34]. The difference between the two peaks is around 19 eV, which is in good agreement with the value reported in the literature [35]. Cl 2p peaks can be deconvoluted into two main doublets, which are found at 200.6 eV and 198.9 eV and correspond to Cl $2p_{1/2}$ and Cl $2p_{3/2}$, respectively. The peak difference of the two peaks is around 1.7 eV, which is in good agreement with the value reported in the literature [36]. The XPS results and the presence of Cu 2p and Cl 2p binding energies confirm the presence of $CuCl_2$ and the effective p-type doping on MoS_2.

To understand the effect of $CuCl_2$ doping on the electrical properties of FETs based on a 2D MoS_2 channel, the electrical properties were measured before and after doping the MoS_2 FETs with $CuCl_2$. The FET devices were fabricated on a HfO_2/Si substrate using photolithography and metal deposition using a thermal evaporator. Figure 4a shows the schematic description of the doping process of MoS_2 FETs. The as-fabricated MoS_2 FETs were spin-coated with $CuCl_2$ solution. Figure 4b shows the representative transfer curve, drain-source current, I_{DS}, as a function of the gate voltage, V_G, which is plotted on a logarithmic scale at the applied drain voltage of V_{DS} = 0.1 V. The inset of Figure 4b shows the transfer curve in a linear scale. The as-fabricated MoS_2 FET showed n-type transfer characteristics and a large ON/OFF ratio above 10^7. From the transfer characteristics, we have estimated a field effect mobility using $\mu_{FE} = \frac{L}{WC_{ox}V_{ds}}\frac{dI_{ds}}{dV_{gs}}$, where L is channel length, W is channel width, and C_{ox} is the gate capacitance of 309.9 nF cm^{-2}. The field effect mobility was measured to be 12.3 cm^2/Vs, which is in good agreement with reported values for a back-gated transistor using CVD-grown MoS_2 monolayers. It can be observed from Figure 4b that the $CuCl_2$-doped MoS_2 FETs shows a gradual decrease in both *ON* and *OFF* current as the $CuCl_2$ is doped onto the MoS_2 FETs.

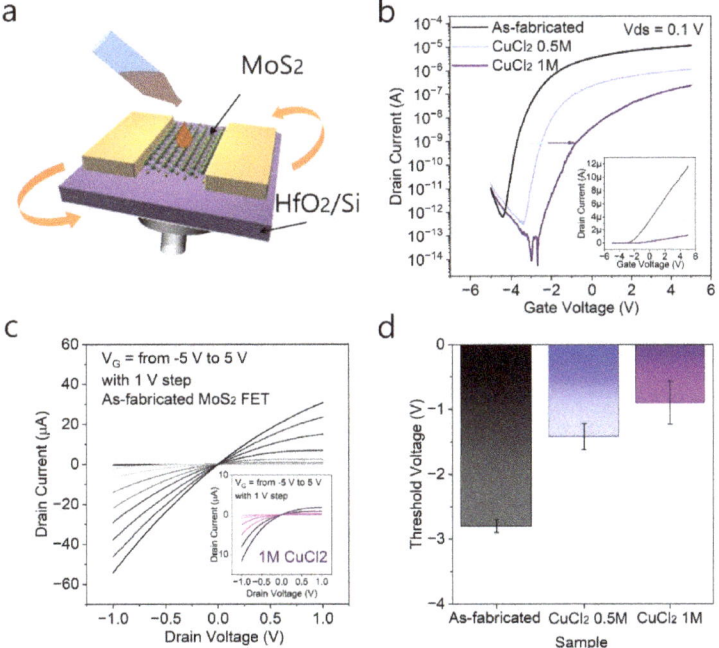

Figure 4. (**a**) Schematic description of $CuCl_2$ doping on MoS_2 FETs. (**b**) The transfer characteristics, (**c**) output characteristics, (**d**) threshold voltages of MoS_2 FETs before and after $CuCl_2$ doping.

Such behavior was also found when an output curve, the drain-source current versus drain-source voltage on a linear scale, was measured as shown in Figure 4c (inset image shows the output curve of 1 M CuCl$_2$ doped MoS$_2$ FET). The output curve of as-fabricated MoS$_2$ FET shows a high ON current and a linearly dependent drain current showing good Ohmic contact between MoS$_2$ and electrodes. As the CuCl$_2$ is doped onto MoS$_2$ FETs, the channel conductance is largely decreased, showing a p-type doping effect of CuCl$_2$ on MoS$_2$, in accordance with transfer curve measured in Figure 4b.

Following the reduced ON current and channel conductance, it was also shown that as CuCl$_2$ was doped onto MoS$_2$ FETs, threshold voltage (V_{th}) was shifted towards positive voltages from -2.85 V to -1.45 V and -0.9 V as 0.5 M and 1 M CuCl$_2$ was doped onto MoS$_2$ FETs as shown in Figure 4d. Using the changes in the V$_{th}$, the change in carrier concentrations upon CuCl$_2$ doping can be calculated using the parallel-plate capacitor model [3,9], $N_{doping} = \frac{C|\Delta V th|}{e}$, where C is the gate capacitance, ΔV_{th} is the change in the V_{Th} after the CuCl$_2$ is doped onto the device compared to the as-fabricated MoS$_2$ FET, and e is the elementary charge. The amount of doping concentration was estimated to be 2.72×10^{12} cm^{-2} and 3.77×10^{12} cm^{-2} when 0.5 M CuCl$_2$ and 1 M CuCl$_2$ were doped onto MoS$_2$ FETs, respectively. The electrical analysis of MoS$_2$ FETs before and after CuCl$_2$ doping show that CuCl$_2$ is an effective p-type dopant for MoS$_2$ that can be easily employed using simple spin coating of different concentrations of CuCl$_2$ solution.

4. Conclusions

To conclude, we have demonstrated simple and effective p-type doping on the 2D MoS$_2$ FETs by simply spin coating the device with CuCl$_2$ solution at ambient conditions. The effect of CuCl$_2$ doping was confirmed analytically through Raman, PL, and XPS measurements. The p-type doping on the MoS$_2$ channel showed largely decreased channel conductance and the shift in threshold voltages towards positive gate voltages in back-gated MoS$_2$ transistors. It was also shown that the amount of doping can be simply controlled by the CuCl$_2$ concentrations in a solution containing ethanol. The results and findings present an important pathway towards designing a CMOS circuit based on 2D FETs and other nanomaterials.

Funding: This work was supported by the Hongik University new faculty research support fund.

Data Availability Statement: Not applicable.

Conflicts of Interest: The author declares no conflict of interest.

References

1. Pak, S.; Lee, J.; Lee, Y.-W.; Jang, A.-R.; Ahn, S.; Ma, K.Y.; Cho, Y.; Hong, J.; Lee, S.; Jeong, H.Y.; et al. Strain-Mediated Interlayer Coupling Effects on the Excitonic Behaviors in an Epitaxially Grown MoS$_2$/WS$_2$ van der Waals Heterobilayer. *Nano Lett.* **2017**, *17*, 5634–5640. [CrossRef] [PubMed]
2. Pak, S.; Cho, Y.; Hong, J.; Lee, J.; Lee, S.; Hou, B.; An, G.H.; Lee, Y.W.; Jang, J.E.; Im, H.; et al. Consecutive Junction-Induced Efficient Charge Separation Mechanisms for High-Performance MoS$_2$/Quantum Dot Phototransistors. *ACS Appl. Mater. Interfaces* **2018**, *10*, 38264–38271. [CrossRef] [PubMed]
3. Pak, S.; Jang, A.R.; Lee, J.; Hong, J.; Giraud, P.; Lee, S.; Cho, Y.; An, G.H.; Lee, Y.W.; Shin, H.S.; et al. Surface functionalization-induced photoresponse characteristics of monolayer MoS$_2$ for fast flexible photodetectors. *Nanoscale* **2019**, *11*, 4726–4734. [CrossRef] [PubMed]
4. Pak, S.; Lee, J.; Jang, A.R.; Kim, S.; Park, K.-H.; Sohn, J.I.; Cha, S. Strain—Engineering of Contact Energy Barriers and Photoresponse Behaviors in Monolayer MoS$_2$ Flexible Devices. *Adv. Funct. Mater.* **2020**, *30*, 2002023. [CrossRef]
5. Pak, S.; Jang, S.; Kim, T.; Lim, J.; Hwang, J.S.; Cho, Y.; Chang, H.; Jang, A.R.; Park, K.H.; Hong, J.; et al. Electrode-Induced Self-Healed Monolayer MoS$_2$ for High Performance Transistors and Phototransistors. *Adv. Mater.* **2021**, *33*, 2102091. [CrossRef]
6. Novoselov, K.S.; Mishchenko, A.; Carvalho, A.; Castro Neto, A.H. 2D materials and van der Waals heterostructures. *Science* **2016**, *353*, aac9439. [CrossRef]
7. Liu, Y.; Weiss, N.O.; Duan, X.; Cheng, H.-C.; Huang, Y.; Duan, X. Van der Waals heterostructures and devices. *Nat. Rev. Mater.* **2016**, *1*, 16042. [CrossRef]
8. Jariwala, D.; Marks, T.J.; Hersam, M.C. Mixed-dimensional van der Waals heterostructures. *Nat. Mater.* **2017**, *16*, 170–181. [CrossRef]

9. Pak, S.; Lim, J.; Hong, J.; Cha, S. Enhanced Hydrogen Evolution Reaction in Surface Functionalized MoS$_2$ Monolayers. *Catalysts* **2021**, *11*, 70. [CrossRef]
10. Liu, Y.; Guo, J.; Zhu, E.B.; Liao, L.; Lee, S.J.; Ding, M.N.; Shakir, I.; Gambin, V.; Huang, Y.; Duan, X.F. Approaching the Schottky-Mott limit in van der Waals metal-semiconductor junctions. *Nature* **2018**, *557*, 696–700. [CrossRef]
11. Wang, Y.; Kim, J.C.; Wu, R.J.; Martinez, J.; Song, X.J.; Yang, J.; Zhao, F.; Mkhoyan, K.A.; Jeong, H.Y.; Chhowalla, M. Van der Waals contacts between three-dimensional metals and two-dimensional semiconductors. *Nature* **2019**, *568*, 70–74. [CrossRef] [PubMed]
12. Shen, P.C.; Su, C.; Lin, Y.X.; Chou, A.S.; Cheng, C.C.; Park, J.H.; Chiu, M.H.; Lu, A.Y.; Tang, H.L.; Tavakoli, M.M.; et al. Ultralow contact resistance between semimetal and monolayer semiconductors. *Nature* **2021**, *593*, 211–217. [CrossRef] [PubMed]
13. Lee, J.; Pak, S.; Lee, Y.-W.; Cho, Y.; Hong, J.; Giraud, P.; Shin, H.S.; Morris, S.M.; Sohn, J.I.; Cha, S.; et al. Monolayer optical memory cells based on artificial trap-mediated charge storage and release. *Nat. Commun.* **2017**, *8*, 14734. [CrossRef] [PubMed]
14. Jung, S.W.; Pak, S.; Lee, S.; Reimers, S.; Mukherjee, S.; Dudin, P.; Kim, T.K.; Cattelan, M.; Fox, N.; Dhesi, S.S.; et al. Spectral functions of CVD grown MoS$_2$ monolayers after chemical transfer onto Au surface. *Appl. Surf. Sci.* **2020**, *532*, 147390. [CrossRef]
15. Liu, H.; Liu, Y.; Zhu, D. Chemical doping of graphene. *J. Mater. Chem.* **2010**, *21*, 3335–3345. [CrossRef]
16. Mouri, S.; Miyauchi, Y.; Matsuda, K. Tunable Photoluminescence of Monolayer MoS$_2$ via Chemical Doping. *Nano Lett.* **2013**, *13*, 5944–5948. [CrossRef]
17. Yang, L.; Majumdar, K.; Liu, H.; Du, Y.; Wu, H.; Hatzistergos, M.; Hung, P.Y.; Tieckelmann, R.; Tsai, W.; Hobbs, C.; et al. Chloride molecular doping technique on 2D materials: WS$_2$ and MoS$_2$. *Nano Lett.* **2014**, *14*, 6275–6280. [CrossRef]
18. Tarasov, A.; Zhang, S.; Tsai, M.-Y.Y.; Campbell, P.M.; Graham, S.; Barlow, S.; Marder, S.R.; Vogel, E.M. Controlled doping of large-area trilayer MoS$_2$ with molecular reductants and oxidants. *Adv. Mater.* **2015**, *27*, 1175–1181. [CrossRef]
19. Cho, Y.; Pak, S.; Li, B.; Hou, B.; Cha, S. Enhanced Direct White Light Emission Efficiency in Quantum Dot Light—Emitting Diodes via Embedded Ferroelectric Islands Structure. *Adv. Funct. Mater.* **2021**, *31*, 2104239. [CrossRef]
20. Kiriya, D.; Tosun, M.; Zhao, P.; Kang, J.S.; Javey, A. Air-stable surface charge transfer doping of MoS$_2$ by benzyl viologen. *J. Am. Chem. Soc.* **2014**, *136*, 7853–7856. [CrossRef]
21. Andleeb, S.; Singh, A.; Eom, J. Chemical doping of MoS$_2$ multilayer by p-toluene sulfonic acid. *Sci. Technol. Adv. Mater.* **2015**, *16*, 035009. [CrossRef] [PubMed]
22. Sim, D.M.; Kim, M.; Yim, S.; Choi, M.J.; Choi, J.; Yoo, S.; Jung, Y.S. Controlled Doping of Vacancy-Containing Few-Layer MoS$_2$ via Highly Stable Thiol-Based Molecular Chemisorption. *ACS Nano* **2015**, *9*, 12115–12123. [CrossRef] [PubMed]
23. Qi, L.; Wang, Y.; Shen, L.; Wu, Y. Chemisorption-induced n-doping of MoS$_2$ by oxygen. *Appl. Phys. Lett.* **2016**, *108*, 063103. [CrossRef]
24. Kim, T.; Pak, S.; Lim, J.; Hwang, J.S.; Park, K.-H.; Kim, B.-S.; Cha, S. Electromagnetic Interference Shielding with 2D Copper Sulfide. *ACS Appl. Mater. Interfaces* **2022**, *14*, 13499–13506. [CrossRef]
25. Zhang, X.K.; Liao, Q.L.; Liu, S.; Kang, Z.; Zhang, Z.; Du, J.L.; Li, F.; Zhang, S.H.; Xiao, J.K.; Liu, B.S.; et al. Poly(4-styrenesulfonate)-induced sulfur vacancy self-healing strategy for monolayer MoS$_2$ homojunction photodiode. *Nat. Commun.* **2017**, *8*, 15881. [CrossRef]
26. Kufer, D.; Konstantatos, G. Highly Sensitive, Encapsulated MoS$_2$ Photodetector with Gate Controllable Gain and Speed. *Nano Lett.* **2015**, *15*, 7307–7313. [CrossRef]
27. Lee, J.; Pak, S.; Giraud, P.; Lee, Y.W.; Cho, Y.; Hong, J.; Jang, A.R.; Chung, H.S.; Hong, W.K.; Jeong, H.; et al. Thermodynamically Stable Synthesis of Large-Scale and Highly Crystalline Transition Metal Dichalcogenide Monolayers and their Unipolar n–n Heterojunction Devices. *Adv. Mater.* **2017**, *29*, 1702206. [CrossRef]
28. Lee, J.; Pak, S.; Lee, Y.W.; Park, Y.; Jang, A.R.; Hong, J.; Cho, Y.; Hou, B.; Lee, S.; Jeong, H.Y.; et al. Direct Epitaxial Synthesis of Selective Two-Dimensional Lateral Heterostructures. *ACS Nano* **2019**, *13*, 13047–13055. [CrossRef]
29. Rybin, M.G.; Islamova, V.R.; Obraztsova, E.A.; Obraztsova, E.D. Modification of graphene electronic properties via controllable gas-phase doping with copper chlroide. *Appl. Phys. Lett.* **2018**, *112*, 033107. [CrossRef]
30. Mansour, A.E.; Kirmani, A.R.; Barlow, S.; Marder, S.R.; Amassian, A. Hybrid Doping of Few-Layer Graphene via a Combination of Intercalation and Surface Doping. *ACS Appl. Mater. Interfaces* **2017**, *9*, 20020–20028. [CrossRef]
31. Kwon, K.C.; Choi, K.S.; Kim, S.Y. Increased Work Function in Few-Layer Graphene Sheets via Metal Chloride Doping. *Adv. Funct. Mater.* **2012**, *22*, 4724–4731. [CrossRef]
32. Chakraborty, B.; Bera, A.; Muthu, D.V.S.; Bhowmick, S.; Waghmare, U.V.; Sood, A.K. Symmetry-dependent phonon renormalization in monolayer MoS$_2$ transistor. *Phys. Rev. B* **2012**, *85*, 161403. [CrossRef]
33. Mak, K.F.; He, K.; Lee, C.; Lee, G.H.; Hone, J.; Heinz, T.F.; Shan, J. Tightly bound trions in monolayer MoS$_2$. *Nat. Mater.* **2013**, *12*, 207–211. [CrossRef] [PubMed]
34. Akgul, F.; Akgul, G.; Yildirim, N.; Unalan, H.; Turan, R. Influence of thermal annealing on microstructural, morphological, optical properties and surface electronic structure of copper oxide thin films. *Mater. Chem. Phys.* **2014**, *147*, 987–995. [CrossRef]
35. Gan, Z.H.; Yu, G.Q.; Tay, B.K.; Tan, C.M.; Zhao, Z.W.; Fu, Y.Q. Preparation and characterization of copper oxide thin films deposited by filtered cathodic vacuum arc. *J. Phys. D Appl. Phys.* **2004**, *37*, 81–85. [CrossRef]
36. Krumpolec, R.; Homola, T.; Cameron, D.; Humlíček, J.; Caha, O.; Kuldová, K.; Zazpe, R.; Přikryl, J.; Macak, J. Structural and Optical Properties of Luminescent Copper(I) Chloride Thin Films Deposited by Sequentially Pulsed Chemical Vapour Deposition. *Coatings* **2018**, *8*, 369. [CrossRef]

Article

Observation of Strong Interlayer Couplings in WS$_2$/MoS$_2$ Heterostructures via Low-Frequency Raman Spectroscopy

Ki Hoon Shin [1,†], Min-Kyu Seo [1,†], Sangyeon Pak [2], A-Rang Jang [3,*] and Jung Inn Sohn [1,*]

1. Division of Physics and Semiconductor Science, Dongguk University, Seoul 04620, Korea; kihoonshin@dongguk.edu (K.H.S.); seominkyuu@gmail.com (M.-K.S.)
2. School of Electronic and Electrical Engineering, Hongik University, Seoul 04066, Korea; spak@hongik.ac.kr
3. Department of Electrical Engineering, Semyung University, Jecheon 27136, Korea
* Correspondence: arjang@semyung.ac.kr (A.-R.J.); junginn.sohn@dongguk.edu (J.I.S.)
† These authors contributed equally to this work.

Abstract: Van der Waals (vdW) heterostructures based on two-dimensional (2D) transition metal dichalcogenides (TMDCs), particularly WS$_2$/MoS$_2$ heterostructures with type-II band alignments, are considered as ideal candidates for future functional optoelectronic applications owing to their efficient exciton dissociation and fast charge transfers. These physical properties of vdW heterostructures are mainly influenced by the interlayer coupling occurring at the interface. However, a comprehensive understanding of the interlayer coupling in vdW heterostructures is still lacking. Here, we present a detailed analysis of the low-frequency (LF) Raman modes, which are sensitive to interlayer coupling, in bilayers of MoS$_2$, WS$_2$, and WS$_2$/MoS$_2$ heterostructures directly grown using chemical vapor deposition to avoid undesirable interfacial contamination and stacking mismatch effects between the monolayers. We clearly observe two distinguishable LF Raman modes, the interlayer in-plane shear and out-of-plane layer-breathing modes, which are dependent on the twisting angles and interface quality between the monolayers, in all the 2D bilayered structures, including the vdW heterostructure. In contrast, LF modes are not observed in the MoS$_2$ and WS$_2$ monolayers. These results indicate that our directly grown 2D bilayered TMDCs with a favorable stacking configuration and high-quality interface can induce strong interlayer couplings, leading to LF Raman modes.

Keywords: bilayer MoS$_2$; bilayer WS$_2$; WS$_2$/MoS$_2$ heterostructure; low-frequency Raman modes; interlayer coupling

1. Introduction

Two-dimensional (2D) transition metal dichalcogenides (TMDCs), such as MoS$_2$ and WS$_2$, have attracted significant attention for flexible and transparent electronics and optoelectronics applications because of their superior mechanical, electrical, and optical properties, such as high mobility, good mechanical strength, superior transparency, and excellent flexibility, as well as strong light–matter interactions [1–6]. TMDCs have considered to be promising nanomaterials for nanoelectronics, optoelectronics, sensors, energy conversion, and energy storage devices over the past decades [7–15]. Recently, considerable efforts have been devoted to the vertical assembling and integration of distinct 2D TMDC monolayers into van der Waals (vdW) heterostructures with diverse band alignments to develop a wide range of optoelectronic devices [16–18]. Because the individual monolayers composing the heterostructure are held together only by weak vdW interactions, the physical properties of such devices are strongly dependent on the interlayer coupling that occurs at the interface of the TMDC-based vdW heterostructures [15,19,20]. Thus, understanding the underlying physics of the interlayer couplings in vdW heterostructures, which play an important role in determining the charge and energy transfer behavior, is crucial for designing and developing high-performance devices.

Among the various heterojunction configurations with the corresponding band structures, the WS_2/MoS_2 heterostructures are considered to be a promising platform for use in device applications to achieve a high device performance because of their type-II heterojunctions, which enable an efficient exciton dissociation and charge transport [21–25]. For instance, Tan et al. demonstrated that the interlayer coupling between WS_2 and MoS_2 can enable a long-lived trap state in WS_2/MoS_2 heterostructures, resulting in an enhanced photodetector performance with a large photoconductive gain and high responsivity [18]. Wang et al. reported that the strong interlayer coupling in WS_2/MoS_2 heterostructures reduces the energy intervals of electron transition, making it detectable under infrared light [19]. Thus, a fast and reliable method for characterizing the interlayer coupling within 2D TMDCs heterostructures is highly desirable for rapidly assessing the physics underlying the unique electronic structures and optical properties.

Raman spectroscopy is one of the most powerful nondestructive tools used for obtaining a detailed structural and electronic information on 2D layered TMDCs and vdW heterostructures [26]. In particular, it has been recently reported that low-frequency (LF) Raman modes can be used as an indicator to directly probe the interlayer coupling effects in vdW-based layered structures, including layered heterostructures, because the LF Raman vibrational modes originate from the weaker interlayer vdW restoring forces [27,28]. In a few studies, the interfaces of WS_2/MoS_2 heterostructures have been investigated using Raman scattering [29,30]. However, a detailed study on the interlayer coupling in WS_2/MoS_2 heterostructures through LF Raman analysis has not yet been conducted.

In this study, we investigated and compared the LF Raman modes in the bilayers of MoS_2, WS_2, and WS_2/MoS_2 heterostructures using a confocal Raman spectrometer to clearly distinguish the LF Raman modes. Moreover, we directly synthesized the MoS_2, WS_2, and WS_2/MoS_2 heterostructures on a SiO_2/Si substrate via chemical vapor deposition (CVD) to rule out the undesirable effects, such as mismatch in the stacking angle between the monolayers and interface contaminations, which generally result in physically stacked TMDCs [27,28,31]. The thicknesses of the directly grown mono- and bilayers were confirmed using atomic force microscopy (AFM). Raman spectroscopy measurements were performed to characterize the vibrational modes in the high-frequency and LF ranges. Two characteristic LF Raman modes, the interlayer in-plane shear (C) and out-of-plane layer-breathing (LB) modes, were clearly observed in all the 2D bilayered homo- and heterostructures, including the MoS_2 and WS_2 bilayers, and WS_2/MoS_2 heterostructures, whereas no LF modes were detected in the MoS_2 and WS_2 monolayers. These results indicate that the LF modes can be ascribed to the strong coupling in our samples without interface contaminations/defects and/or misalignment effects, compared to mechanically stacked samples.

2. Materials and Methods

2.1. Sample Preparation

First, a solution was prepared by adding MoO_3 powder (200 mg, Sigma-Aldrich, St. Louis, MO, USA) and WO_3 powder (200 mg, Sigma-Aldrich, St. Louis, MO, USA) to NH_4OH (10 mL, 28–30% solution, Sigma-Aldrich, St. Louis, MO, USA) in a small vial. A 20 µL MoO_3 and WO_3 solution was dropped onto a substrate and spin coated at 3000 rpm for 1 min. The weight of the deposited MoO_3 film and WO_3 particles was found to be significantly small, around ~0.01 mg [32].

Bilayer MoS_2 were grown using atmospheric pressure CVD (APCVD) (Figure 1a). The prepared MoO_3 solution was coated on a 300-nm-thick SiO_2/Si substrate and loaded into the center of a quartz tube furnace. A ceramic boat loaded with 100 mg of S powder was placed upstream of the furnace. The bilayer MoS_2 growth was carried out at 730 °C for 10 min, and then the furnace was naturally cooled down to room temperature.

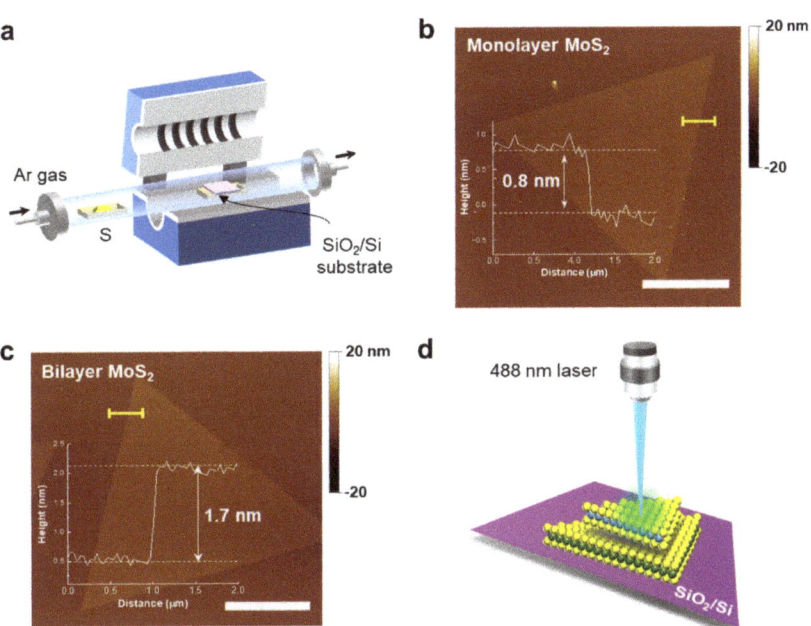

Figure 1. (**a**) Schematic of the growth process of the 2D bilayered MoS_2, WS_2, and WS_2/MoS_2 heterostructures via CVD. AFM topography images of the directly grown (**b**) monolayer and (**c**) bilayer MoS_2. (Scale bar: 5 μm) (**d**) Schematic of the Raman spectroscopy measurements with a 488 nm laser focused on the vdW heterostructures.

Bilayer WS_2 were grown using low-pressure CVD (LPCVD), in which the quartz tube was evacuated to a base pressure of approximately 10^{-3} Torr, and a mixture of Ar and H_2 was flowed into the furnace. The prepared WO_3 solution and S powder were placed in the same position as that fixed during the bilayer MoS_2 growth process. The bilayer WS_2 growth was performed at 850 °C for 10 min, and throughout the process, the furnace pressure was maintained at 15 Torr.

Next, a WS_2/MoS_2 heterostructure was grown using the LPCVD method. The substrates coated with the MoO_3 and WO_3 solution were loaded on the bottom and top of the crucible in a furnace, respectively. The S powder was placed in the same position as that fixed during the bilayer MoS_2 growth process, and the growth temperature, time, and pressure were the same as those used in the bilayer WS_2 growth process. As the temperature of the furnace increased, MoO_3 was deposited and sulfurized on the substrate located at the top of the crucible. The temperature of furnace reached at 850 °C, and the vertical and lateral MoS_2/WS_2 heterostructures were synthesized [32].

2.2. Characterization

The surface morphology and thicknesses of the CVD-grown bilayer TMDs were characterized using atomic force microscopy (AFM (Multimode 8, Bruker, Billerica, MA, USA)).

Observing the LF Raman modes in 2D layered TMDCs and vdW heterostructures is challenging, because it is difficult to distinguish them from the Rayleigh line using a conventional Raman microscope [33]. In this study, Raman spectroscopy measurements were performed at room temperature using a confocal Raman spectrometer (alpha 300 M+, WiTech, Ulm, Germany). A green laser beam with a wavelength of 488 nm and power of approximately 1 mW was focused onto the individual samples by a 50× objective (NA:

0.55) with a long working distance, as illustrated in Figure 1d. The estimated laser spot size was 1.08 µm.

3. Results and Discussion

3.1. MoS$_2$ Bilayers

Monolayer and bilayer MoS$_2$ were directly grown on a SiO$_2$/Si substrate via the APCVD method, as shown in Figure 1a (see the Section 2 for more details).

Figure 1b,c show the AFM images of the triangular mono- and bilayer MoS$_2$. The lateral size of both the layers was approximately 13–15 µm. The height profiles show that the thicknesses of the mono- and bilayer MoS$_2$ were found to be approximately 0.8 and 1.7 nm, respectively.

Raman spectroscopy was performed to investigate the vibrational modes of the as-grown 2D bilayer MoS$_2$, WS$_2$, and WS$_2$/MoS$_2$ heterostructures. Figure 2a illustrates the four different vibrational modes with relative displacements between the metal and chalcogen atoms, which can be representatively observed in most few-layer TMDCs. These modes can be classified into two categories according to their frequencies: high-frequency and LF modes. In the high-frequency range (>300 cm^{-1}), the E$_{2g}$ and A$_{1g}$ modes originate from the in-plane and out-of-plane atomic vibrations within the layers, respectively. It was observed that the E$_{2g}$ mode softens (red-shift), while the A$_{1g}$ mode stiffens (blue-shift) as the number of layers is increased. Consequently, the frequency difference between the two modes increases [34]. Because the high-frequency Raman modes of the mono- or multilayered TMDCs originate from the intralayer chemical bonds, the restoring forces are significantly affected by the strength of the intralayer bonding, resulting in less sensitivity to interlayer coupling [26,27,34,35]. In contrast, in the LF range (<50 cm^{-1}), the interlayer in-plane shear (C) and out-of-plane layer-breathing (LB) modes are expected to have very low frequencies owing to the weak interlayer vdW restoring forces [36,37]. Thus, the analysis of the LF Raman modes can be treated as a more reliable method for directly probing the interlayer coupling in multilayered TMDCs.

Figure 2b shows the Raman spectra of the monolayer (black line) and bilayer MoS$_2$ (red line), recorded in the high-frequency range of 320–480 cm^{-1}. Note that the quality and lateral size of the MoS$_2$ (and WS$_2$) samples chosen for the Raman measurement were almost the same as those shown in Figure 1b,c. For the monolayer MoS$_2$, two typical peaks were obtained at 386.7 and 404.7 cm^{-1}, which correspond to the in-plane vibrational mode (E$_{2g}$ mode) and out-of-plane vibrational mode (A$_{1g}$ mode), respectively. Evidently, for the bilayer MoS$_2$, the E$_{2g}$ and A$_{1g}$ modes shift in the opposite direction, and the frequency difference between these two modes increases from 18.1 cm^{-1} (for the monolayer MoS$_2$) to 21.1 cm^{-1} [34,38,39].

Figure 2c shows the Raman spectra of the monolayer (black line) and bilayer MoS$_2$ (red line), obtained in the LF range of −80 to 80 cm^{-1}. Neither the C nor the LB modes are observed in the monolayer MoS$_2$, as is generally expected for all the other monolayer TMDCs, because the interlayer restoring force is absent. In contrast to the monolayer MoS$_2$, two distinct LF Raman peaks were observed in the spectra of the bilayer MoS$_2$, as well as in the corresponding anti-Stokes spectra. A sharp peak at 40.2 cm^{-1} can be assigned to the LB mode of the bilayer MoS$_2$. Another peak at 24.3 cm^{-1} can be assigned to the C mode of the bilayer MoS$_2$, which is typically of lower frequency than that of the LB mode.

According to previous studies, the intensity and frequency of the Raman peaks corresponding to the C and LB modes in bilayer MoS$_2$ depend on the twisting angles and interface quality between the top and bottom MoS$_2$ layers. It has been reported that the C mode is clearly observed near 0° or 60°, and is even absent for certain twisting angles. [31] Thus, the clear observation of the C mode shown in Figure 2c implies that the bilayer MoS$_2$ was grown with the 2H (60°) or 3R (0°) stacking configuration (as shown in Figure 1b,c), which is the most stable configuration of CVD-grown bilayer systems [40]. Furthermore, we observed that there exists only one peak corresponding to the LB mode. This finding indicates that unlike the mechanically stacked bilayer MoS$_2$, the directly grown bilayer

MoS$_2$ had a uniform interlayer without localized strains, crystallographic defects, or wrinkles, which are usually introduced in mechanically transferred samples. Thus, these results suggest that our directly grown bilayer MoS$_2$ can induce a strong interlayer coupling with highly ordered domains between the monolayers.

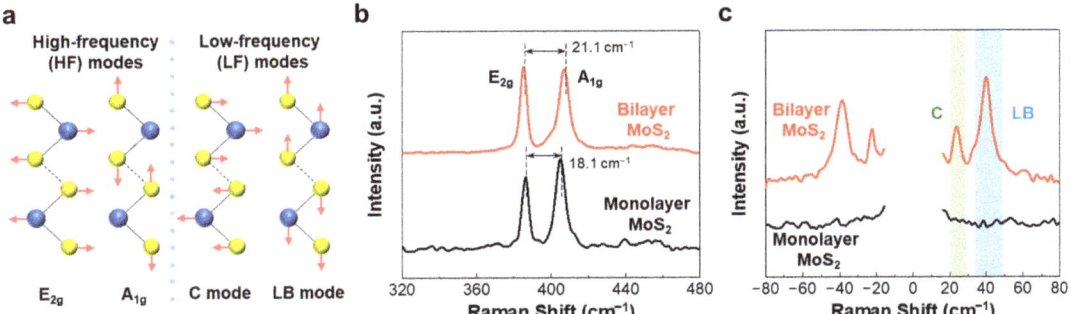

Figure 2. (**a**) Schematic of the lattice structure and four different vibrational modes of the bilayered TMDCs. Raman spectra of the monolayer (black lines) and bilayer MoS$_2$ (red lines), obtained in the (**b**) high-frequency and (**c**) LF ranges. The Raman peaks enclosed by the green- and blue-colored regions correspond to the C and LB modes, respectively.

3.2. WS$_2$ Bilayers

Monolayer and bilayer WS$_2$ were directly grown on a SiO$_2$/Si substrate via an LPCVD method (see Section 2 for more details).

Figure 3a shows the Raman spectra of the monolayer (black line) and bilayer WS$_2$ (red line), obtained in the high-frequency range of 320–480 cm^{-1}. For the monolayer WS$_2$, two typical Raman signals were observed at 356.5 and 417.5 cm^{-1}, corresponding to the in-plane vibrational mode (E$_{2g}$ mode) and out-of-plane vibrational mode (A$_{1g}$ mode) of WS$_2$, respectively. Moreover, the E$_{2g}$ and A$_{1g}$ modes of the bilayer WS$_2$ shifted in the opposite direction compared to those of the monolayer WS$_2$, resulting in an increase in the frequency difference between them, which is consistent with the previously reported results [41,42].

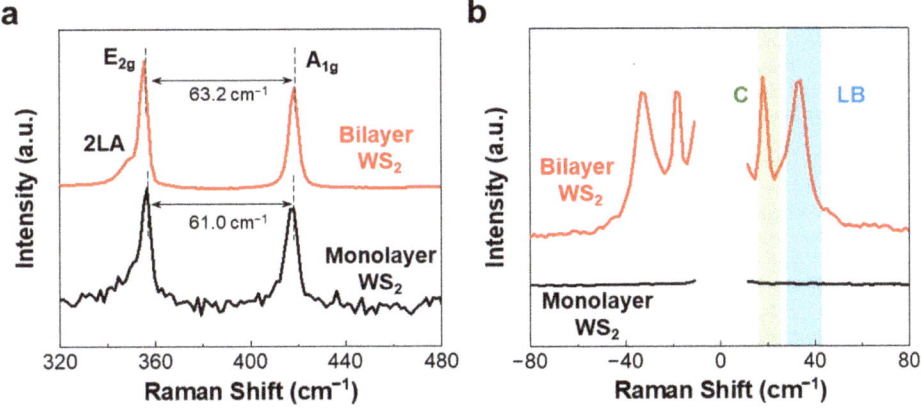

Figure 3. Raman spectra of the mono- (black lines) and bilayer WS$_2$ (red lines), recorded in the (**a**) high-frequency and (**b**) LF ranges. The Raman peaks enclosed by the green- and blue-colored regions correspond to the C and LB modes, respectively.

Figure 3b shows the Raman spectra of the monolayer (black line) and bilayer WS_2 (red line), obtained in the LF range of −80 to 80 cm^{-1}. Similar to the results of MoS_2 shown in Figure 2c, for the monolayer WS_2, no peaks corresponding to the C and LB modes were observed. However, two different peaks at 17.9, and 33.8 cm^{-1} were observed in the LF Raman spectrum of the bilayer WS_2, which can be assigned to the C and LB modes in a Stokes Raman spectrum, respectively. These findings indicate that the LF Raman modes can be attributed to the interlayer coupling effect.

3.3. WS_2/MoS_2 Heterostructures

A vertical WS_2/MoS_2 heterostructure, with WS_2 on top of the MoS_2 monolayer, was directly grown on a 300-nm-thick SiO_2/Si substrate via the LPCVD method (see Section 2 for more details).

The Raman spectrum of the WS_2/MoS_2 heterostructure in the high-frequency range of 320–500 cm^{-1} is presented in Figure 4a. The characteristic peaks observed at 383.7 and 356.5 cm^{-1} can be ascribed to the E_{2g} modes of the individual MoS_2 and WS_2 monolayers, respectively, whereas those located at 404.7 and 418.2 cm^{-1} can be attributed to the A_{1g} modes of these individual MoS_2 and WS_2 monolayers, respectively, confirming the formation of the WS_2/MoS_2 heterostructure.

In the LF range, as shown in Figure 4b, we also observed two sharp characteristic peaks at 19.5 and 33.8 cm^{-1}, which can be assigned to the C and LB modes, respectively. Figure 4c shows the LF modes of the WS_2/MoS_2 heterostructure. In general, two physically transferred layers composed of vdW heterostructures are usually misaligned, leading to a lack of in-plane restoring force, thus resulting in the disappearance of the C mode. Furthermore, nonuniform interfaces with variable local stacking can result in the presence of multiple LB modes [27]. Unlike the observations made in a previous study [29], here, we observed that in the WS_2/MoS_2 heterostructure, only one peak existed corresponding to each C and LB mode, which were associated with a good stacking configuration and interface uniformity, respectively [31]. Notably, directly grown heterostructures with epitaxial interfaces can exhibit a strong interlayer coupling compared to the mechanically transferred heterostructures prepared using a dry-transfer method or exfoliated suspension drop casting [28,43]. Thus, we believe that the clear observation of the C and LB mode peaks can be attributed to the strong interlayer coupling arising from the epitaxially grown bilayers with a clear interface and stable configuration.

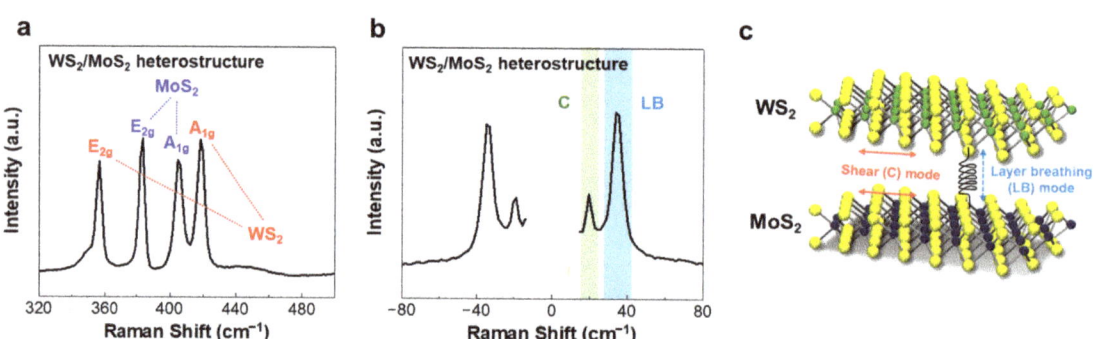

Figure 4. Raman spectra of the WS_2/MoS_2 heterostructures recorded in the (**a**) high-frequency and (**b**) LF ranges. The Raman peaks enclosed by the green- and blue-colored regions correspond to the C and LB modes, respectively. (**c**) Schematic depicting the interlayer interaction categorized by the C and LB modes in the LF range for the WS_2/MoS_2 heterostructure.

4. Conclusions

In this study, we investigated the LF Raman modes to explore the interlayer coupling in the bilayered MoS_2 structures, WS_2 structures, and WS_2/MoS_2 heterostructures, which were directly synthesized via CVD. For all the 2D bilayered MoS$_2$ and WS$_2$ homostructures and WS$_2$/MoS$_2$ heterostructures, typical LF C and LB modes were observed, whereas no LB modes were detected in the MoS$_2$ and WS$_2$ monolayers. Moreover, we showed that the observed single LB mode and clear C mode peaks could be attributed to the high-quality homo- and heterojunctions with stable stacking configurations, which enable the induction of a strong interlayer coupling within the layers. Our results provide a fundamental understanding of the interlayer coupling in WS$_2$/MoS$_2$ heterostructures and other 2D TMDC-based vdW heterostructures and an ideal approach for designing and developing high-performance functional devices based on various vdW heterostructures.

Author Contributions: Conceptualization, A.-R.J. and J.I.S.; methodology, K.H.S. and M.-K.S.; validation, A.-R.J. and J.I.S.; formal analysis S.P. and A.-R.J.; investigation, K.H.S., M.-K.S. and S.P.; resources, A.-R.J. and J.I.S.; data curation, K.H.S., M.-K.S., A.-R.J. and J.I.S.; writing—original draft preparation, K.H.S. and M.-K.S.; writing—review and editing, S.P., A.-R.J. and J.I.S.; visualization, K.H.S.; supervision, J.I.S.; funding acquisition, A.-R.J. and J.I.S. All authors have read and agreed to the published version of the manuscript.

Funding: This research was supported by the National Research Foundation of Korea (NRF) grant funded by the Korean government (MSIT) (2019R1A2C1007883 and 2021R1I1A3049729). This work was also supported by 2022 Hongik University Research Fund.

Data Availability Statement: Not applicable.

Conflicts of Interest: The authors declare no conflict of interest.

References

1. Manzeli, S.; Ovchinnikov, D.; Pasquier, D.; Yazyev, O.V.; Kis, A. 2D transition metal dichalcogenides. *Nat. Rev. Mater.* **2017**, *2*, 17033. [CrossRef]
2. Fuhrer, M.S.; Hone, J. Measurement of mobility in dual-gated MoS$_2$ Transistors. *Nat. Nanotechnol.* **2013**, *8*, 146–147. [CrossRef] [PubMed]
3. Lee, J.; Pak, S.; Lee, Y.W.; Cho, Y.; Hong, J.; Giraud, P.; Shin, H.S.; Morris, S.M.; Sohn, J.I.; Cha, S.N.; et al. Monolayer optical memory cells based on artificial trap-mediated charge storage and release. *Nat. Commun.* **2017**, *8*, 14734. [CrossRef] [PubMed]
4. Choi, W.; Choudhary, N.; Han, G.H.; Park, J.; Akinwande, D.; Lee, Y.H. Recent development of two-dimensional transition metal dichalcogenides and their applications. *Mater. Today* **2017**, *20*, 116–130. [CrossRef]
5. Liu, X.; Galfsky, T.; Sun, Z.; Xia, F.; Lin, E.C.; Lee, Y.H.; Kéna-Cohen, S.; Menon, V.M. Strong light-matter coupling in two-dimensional atomic crystals. *Nat. Photonics* **2014**, *9*, 30–34. [CrossRef]
6. Pak, S.; Lee, J.; Jang, A.-R.; Kim, S.; Park, K.-H.; Sohn, J.I.; Cha, S. Strain-engineering of contact energy barriers and photoresponse behaviors in monolayer. *Adv. Funct. Mater.* **2020**, *30*, 2002023. [CrossRef]
7. van der Zande, A.M.; Huang, P.Y.; Chenet, D.A.; Berkelbach, T.C.; You, Y.; Lee, G.H.; Heinz, T.F.; Reichman, D.R.; Muller, D.A.; Hone, J.C. Grains and grain boundaries in highly crystalline monolayer molybdenum disulphide. *Nat. Mater.* **2013**, *12*, 554–561. [CrossRef]
8. Bilgin, I.; Liu, F.; Vargas, A.; Winchester, A.; Man, M.K.L.; Upmanyu, M.; Dani, K.M.; Gupta, G.; Talapatra, S.; Mohite, A.D.; et al. Chemical vapor deposition synthesized atomically thin molybdenum disulfide with optoelectronic-grade crystalline quality. *ACS Nano* **2015**, *9*, 8822–8832. [CrossRef]
9. Voiry, D.; Yamaguchi, H.; Li, J.; Silva, R.; Alves, D.C.B.; Fujita, T.; Chen, M.; Asefa, T.; Shenoy, V.B.; Eda, G.; et al. Enhanced catalytic activity in strained chemically exfoliated WS$_2$ nanosheets for hydrogen evolution. *Nat. Mater.* **2013**, *12*, 850–855. [CrossRef]
10. Lee, D.; Jang, A.R.; Kim, J.Y.; Lee, G.; Jung, D.W.; Lee, T.I.; Lee, J.O.; Kim, J.J. Phase-dependent gas sensitivity of MoS$_2$ chemical sensors investigated with phase-locked MoS$_2$. *Nanotechnology* **2020**, *31*, 225504. [CrossRef]
11. Huang, H.; Cui, Y.; Li, Q.; Dun, C.; Zhou, W.; Huang, W.; Chen, L.; Hewitt, C.A.; Carroll, D.L. Metallic 1T phase MoS$_2$ nanosheets for high-performance thermoelectric energy harvesting. *Nano Energy* **2016**, *26*, 172–179. [CrossRef]
12. Yu, X.; Chen, X.; Ding, X.; Yu, X.; Zhao, X.; Chen, X. Facile fabrication of flower-like MoS$_2$/nanodiamond nanocomposite toward high-performance humidity detection. *Sens. Actuators B Chem.* **2020**, *317*, 128168. [CrossRef]
13. Zhao, Z.; Hu, Z.; Li, Q.; Li, H.; Zhang, X.; Zhuang, Y.; Wang, F.; Yu, G. Designing two-dimensional WS$_2$ layered cathode for high-performance aluminum-ion batteries: From micro-assemblies to insertion mechanism. *Nano Today* **2020**, *32*, 100870. [CrossRef]
14. Lee, J.; Pak, S.; Giraud, P.; Lee, Y.-W.; Cho, Y.; Hong, J.; Jang, A.R.; Chung, H.-S.; Hong, W.-K.; Jeong, H.Y.; et al. Thermodynamically stable synthesis of large-scale and highly crystalline transition metal dichalcogenide monolayers and their unipolar n–n heterojunction devices. *Adv. Mater.* **2017**, *29*, 1702206. [CrossRef]

15. Pak, S.; Lee, J.; Lee, Y.W.; Jang, A.R.; Ahn, S.; Ma, K.Y.; Cho, Y.; Hong, J.; Lee, S.; Jeong, H.Y.; et al. Strain-mediated interlayer coupling effects on the excitonic behaviors in an epitaxially Grown MoS$_2$/WS$_2$ van der Waals heterobilayer. *Nano Lett.* **2017**, *17*, 5634–5640. [CrossRef]
16. Duan, X.; Wang, C.; Pan, A.; Yu, R.; Duan, X. Two-dimensional transition metal dichalcogenides as atomically thin semiconductors: Opportunities and challenges. *Chem. Soc. Rev.* **2015**, *44*, 8859–8876. [CrossRef]
17. Zhu, J.; Li, W.; Huang, R.; Ma, L.; Sun, H.; Choi, J.H.; Zhang, L.; Cui, Y.; Zou, G. One-pot selective epitaxial growth of large WS$_2$/MoS$_2$ lateral and vertical heterostructures. *J. Am. Chem. Soc.* **2020**, *142*, 16276–16284. [CrossRef]
18. Wang, S.; Cui, X.; Jian, C.; Cheng, H.; Niu, M.; Yu, J.; Yan, J.; Huang, W. Stacking engineered heterostructures in transition metal dichalcogenides. *Adv. Mater.* **2021**, *33*, 2005735. [CrossRef]
19. Hong, X.; Kim, J.; Shi, S.F.; Zhang, Y.; Jin, C.; Sun, Y.; Tongay, S.; Wu, J.; Zhang, Y.; Wang, F. Ultrafast charge transfer in atomically thin MoS$_2$/WS$_2$ heterostructures. *Nat. Nanotechnol.* **2014**, *9*, 682–686. [CrossRef]
20. Wu, X.; Chen, X.; Yang, R.; Zhan, J.; Ren, Y.; Li, K. Recent Advances on tuning the interlayer coupling and properties in van Der Waals heterostructures. *Small* **2022**, *18*, 2105877. [CrossRef]
21. Gong, Y.; Lin, J.; Wang, X.; Shi, G.; Lei, S.; Lin, Z.; Zou, X.; Ye, G.; Vajtai, R.; Yakobson, B.I.; et al. Vertical and in-plane heterostructures from WS$_2$/MoS$_2$ monolayers. *Nat. Mater.* **2014**, *13*, 1135–1142. [CrossRef] [PubMed]
22. Yan, J.; Ma, C.; Huang, Y.; Yang, G. Tunable control of interlayer excitons in WS$_2$/MoS$_2$ heterostructures via strong coupling with enhanced Mie resonances. *Adv. Sci.* **2019**, *6*, 1802092. [CrossRef] [PubMed]
23. Susarla, S.; Manimunda, P.; Morais Jaques, Y.; Hachtel, J.A.; Idrobo, J.C.; Syed Amnulla, S.A.; Galvão, D.S.; Tiwary, C.S.; Ajayan, P.M. Deformation mechanisms of vertically stacked WS$_2$/MoS$_2$ heterostructures: The role of interfaces. *ACS Nano* **2018**, *12*, 4036–4044. [CrossRef] [PubMed]
24. Tan, H.; Xu, W.; Sheng, Y.; Lau, C.S.; Fan, Y.; Chen, Q.; Tweedie, M.; Wang, X.; Zhou, Y.; Warner, J.H. Lateral graphene-contacted vertically stacked WS$_2$/MoS$_2$ hybrid photodetectors with large Gain. *Adv. Mater.* **2017**, *29*, 1702917. [CrossRef] [PubMed]
25. Wang, G.; Li, L.; Fan, W.; Wang, R.; Zhou, S.; Lü, J.T.; Gan, L.; Zhai, T. Interlayer coupling induced infrared response in WS$_2$/MoS$_2$ heterostructures enhanced by surface plasmon resonance. *Adv. Funct. Mater.* **2018**, *28*, 1800339. [CrossRef]
26. Zhang, X.; Qiao, X.F.; Shi, W.; Wu, J.B.; Jiang, D.S.; Tan, P.H. Phonon and Raman scattering of two-dimensional transition metal dichalcogenides from monolayer, multilayer to bulk material. *Chem. Soc. Rev.* **2015**, *44*, 2757–2785. [CrossRef]
27. Liang, L.; Zhang, J.; Sumpter, B.G.; Tan, Q.H.; Tan, P.H.; Meunier, V. Low-frequency shear and layer-breathing modes in Raman scattering of two-dimensional materials. *ACS Nano* **2017**, *11*, 11777–11802. [CrossRef]
28. Zhang, J.; Wang, J.H.; Chen, P.; Sun, Y.; Wu, S.; Jia, Z.Y.; Lu, X.B.; Yu, H.; Chen, W.; Zhu, J.Q.; et al. Observation of strong interlayer coupling in MoS$_2$/WS$_2$ heterostructures. *Adv. Mater.* **2016**, *28*, 1950–1956. [CrossRef]
29. Zhang, F.; Lu, Z.; Choi, Y.; Liu, H.; Zheng, H.; Xie, L.; Park, K.; Jiao, L.; Tao, C. Atomically resolved observation of continuous interfaces between an as-grown MoS$_2$ monolayer and a WS$_2$/MoS$_2$ heterobilayer on SiO$_2$. *ACS Appl. Nano Mater.* **2018**, *1*, 2041–2048. [CrossRef]
30. Saito, Y.; Kondo, T.; Ito, H.; Okada, M.; Shimizu, T.; Kubo, T.; Kitaura, R. Low frequency Raman study of interlayer couplings in WS$_2$-MoS$_2$ van der Waals heterostructures. *Jpn. J. Appl. Phys.* **2020**, *59*, 062004. [CrossRef]
31. Huang, S.; Liang, L.; Ling, X.; Puretzky, A.A.; Geohegan, D.B.; Sumpter, B.G.; Kong, J.; Meunier, V.; Dresselhaus, M.S. Low-Frequency Interlayer Raman modes to probe interface of twisted bilayer MoS$_2$. *Nano Lett.* **2016**, *16*, 1435–1444. [CrossRef] [PubMed]
32. Lee, J.; Pak, S.; Lee, Y.W.; Park, Y.; Jang, A.R.; Hong, J.; Cho, Y.; Hou, B.; Lee, S.; Jeong, H.Y.; et al. Direct Epitaxial Synthesis of Selective Two-Dimensional Lateral Heterostructures. *ACS Nano* **2019**, *13*, 13047–13055. [CrossRef] [PubMed]
33. Zhao, Y.; Luo, X.; Li, H.; Zhang, J.; Araujo, P.T.; Gan, C.K.; Wu, J.; Zhang, H.; Quek, S.Y.; Dresselhaus, M.S.; et al. Interlayer breathing and shear modes in few-trilayer MoS$_2$ and WSe$_2$. *Nano Lett.* **2013**, *13*, 1007–1015. [CrossRef] [PubMed]
34. Lee, C.; Yan, H.; Brus, L.E.; Heinz, T.F.; Hone, J.; Ryu, S. Anomalous lattice vibrations of single- and few-layer MoS$_2$. *ACS Nano* **2010**, *4*, 2695–2700. [CrossRef] [PubMed]
35. Sun, Y.; Cai, D.; Yang, Z.; Li, H.; Li, Q.; Jia, D.; Zhou, Y. The Preparation, microstructure and mechanical properties of a dense MgO-Al$_2$O$_3$-SiO$_2$ based glass-ceramic coating on porous BN/Si$_2$N$_2$O ceramics. *RSC Adv.* **2018**, *8*, 17569–17574. [CrossRef]
36. Zeng, H.; Zhu, B.; Liu, K.; Fan, J.; Cui, X.; Zhang, Q.M. Low-frequency Raman modes and electronic excitations in atomically thin MoS$_2$ films. *Phys. Rev. B* **2012**, *86*, 241301. [CrossRef]
37. Sam, R.T.; Umakoshi, T.; Verma, P. Probing stacking configurations in a few layered MoS$_2$ by low frequency Raman spectroscopy. *Sci. Rep.* **2020**, *10*, 21227. [CrossRef]
38. Li, H.; Zhang, Q.; Yap, C.C.R.; Tay, B.K.; Edwin, T.H.T.; Olivier, A.; Baillargeat, D. From bulk to monolayer MoS$_2$: Evolution of Raman scattering. *Adv. Funct. Mater.* **2012**, *22*, 1385–1390. [CrossRef]
39. Najmaei, S.; Liu, Z.; Ajayan, P.M.; Lou, J. Thermal effects on the characteristic Raman spectrum of molybdenum disulfide (MoS$_2$) of varying thicknesses. *Appl. Phys. Lett.* **2012**, *100*, 013106. [CrossRef]
40. Puretzky, A.A.; Liang, L.; Li, X.; Xiao, K.; Wang, K.; Mahjouri-Samani, M.; Basile, L.; Idrobo, J.C.; Sumpter, B.G.; Meunier, V.; et al. Low-frequency Raman fingerprints of two-dimensional metal dichalcogenide layer stacking configurations. *ACS Nano* **2015**, *9*, 6333–6342. [CrossRef]

1. Berkdemir, A.; Gutiérrez, H.R.; Botello-Méndez, A.R.; Perea-López, N.; Elías, A.L.; Chia, C.I.; Wang, B.; Crespi, V.H.; López-Urías, F.; Charlier, J.C.; et al. Identification of individual and few layers of WS_2 using Raman spectroscopy. *Sci. Rep.* **2013**, *3*, 1755. [CrossRef]
2. Zhao, W.; Ghorannevis, Z.; Amara, K.K.; Pang, J.R.; Toh, M.; Zhang, X.; Kloc, C.; Tan, P.H.; Eda, G. Lattice dynamics in mono- and few-layer sheets of WS_2 and WSe_2. *Nanoscale* **2013**, *5*, 9677–9683. [CrossRef] [PubMed]
3. Samad, L.; Bladow, S.M.; Ding, Q.; Zhuo, J.; Jacobberger, R.M.; Arnold, M.S.; Jin, S. Layer-controlled chemical vapor deposition growth of MoS_2 vertical heterostructures via van der Waals epitaxy. *ACS Nano* **2016**, *10*, 7039–7046. [CrossRef] [PubMed]

Article

Optical Mode Tuning of Monolayer Tungsten Diselenide (WSe$_2$) by Integrating with One-Dimensional Photonic Crystal through Exciton–Photon Coupling

Konthoujam James Singh [1], Hao-Hsuan Ciou [1,2], Ya-Hui Chang [1,2], Yen-Shou Lin [1,2], Hsiang-Ting Lin [2], Po-Cheng Tsai [2], Shih-Yen Lin [2], Min-Hsiung Shih [1,2,3,*] and Hao-Chung Kuo [1,2,*]

[1] Department of Photonics, Institute of Electro-Optical Engineering, College of Electrical and Computer Engineering, National Yang Ming Chiao Tung University, Hsinchu 30010, Taiwan; jamesk231996@gmail.com (K.J.S.); tony51415@gmail.com (H.-H.C.); dnn1227@gmail.com (Y.-H.C.); zxc35789512@gmail.com (Y.-S.L.)
[2] Research Center for Applied Sciences (RCAS), Academia Sinica, Taipei 11529, Taiwan; linst8168@gmail.com (H.-T.L.); d07943004@ntu.edu.tw (P.-C.T.); shiyen@gate.sinica.edu.tw (S.-Y.L.)
[3] Department of Photonics, National Sun Yat-sen University, Kaohsiung 80424, Taiwan
* Correspondence: mhshih@gate.sinica.edu.tw (M.-H.S.); hckuo@faculty.nctu.edu.tw (H.-C.K.); Tel.: +886-3-5712121 (H.-C.K.)

Citation: James Singh, K.; Ciou, H.-H.; Chang, Y.-H.; Lin, Y.-S.; Lin, H.-T.; Tsai, P.-C.; Lin, S.-Y.; Shih, M.-H.; Kuo, H.-C. Optical Mode Tuning of Monolayer Tungsten Diselenide (WSe$_2$) by Integrating with One-Dimensional Photonic Crystal through Exciton–Photon Coupling. *Nanomaterials* 2022, *12*, 425. https://doi.org/10.3390/nano12030425

Academic Editors: Jung-Inn Sohn and Sangyeon Pak

Received: 30 December 2021
Accepted: 26 January 2022
Published: 27 January 2022

Publisher's Note: MDPI stays neutral with regard to jurisdictional claims in published maps and institutional affiliations.

Copyright: © 2022 by the authors. Licensee MDPI, Basel, Switzerland. This article is an open access article distributed under the terms and conditions of the Creative Commons Attribution (CC BY) license (https://creativecommons.org/licenses/by/4.0/).

Abstract: Two-dimensional materials, such as transition metal dichalogenides (TMDs), are emerging materials for optoelectronic applications due to their exceptional light–matter interaction characteristics. At room temperature, the coupling of excitons in monolayer TMDs with light opens up promising possibilities for realistic electronics. Controlling light–matter interactions could open up new possibilities for a variety of applications, and it could become a primary focus for mainstream nanophotonics. In this paper, we show how coupling can be achieved between excitons in the tungsten diselenide (WSe$_2$) monolayer with band-edge resonance of one-dimensional (1-D) photonic crystal at room temperature. We achieved a Rabi splitting of 25.0 meV for the coupled system, indicating that the excitons in WSe$_2$ and photons in 1-D photonic crystal were coupled successfully. In addition to this, controlling circularly polarized (CP) states of light is also important for the development of various applications in displays, quantum communications, polarization-tunable photon source, etc. TMDs are excellent chiroptical materials for CP photon emitters because of their intrinsic circular polarized light emissions. In this paper, we also demonstrate that integration between the TMDs and photonic crystal could help to manipulate the circular dichroism and hence the CP light emissions by enhancing the light–mater interaction. The degree of polarization of WSe$_2$ was significantly enhanced through the coupling between excitons in WSe$_2$ and the PhC resonant cavity mode. This coupled system could be used as a platform for manipulating polarized light states, which might be useful in optical information technology, chip-scale biosensing and various opto-valleytronic devices based on 2-D materials.

Keywords: transition metal dichalogenides; excitons; light–matter interactions

1. Introduction

Controlling the light–matter interaction on subwavelength scales is vital for a multitude of nanotechnology applications, such as modulators, lasers, switches, waveguides, logic elements, etc. Two-dimensional (2-D) materials, such as transition metal dichalcogenides (TMDs), have attracted substantial interest in photonics with the advancement of science and technology owing to its remarkable features, such as strong excitonic effects and valley-dependent characteristics [1,2]. It is possible to control the spin and valley in monolayer TMDs due to the strong spin–orbit coupling and breaking of inversion symmetry, which is different from their bulk counterparts [3,4]. In contrast to graphene, the strong quantum confinement in the out-of-plane direction causes the bandgap of TMDs to be strongly influenced by the number of layers; the bandgap can be tuned from indirect to direct bandgap as the number of layers decreases from a few layers to a monolayer [5]. Because of their distinctive direct band gap structure, significant exciton-binding energies of a few 100 meV, and valley-associated features, such as valleytronics, these atomically thin monolayer TMDs have gained a lot of attention. The low dimensionality features of TMDs lead to an efficient light absorption and strong light–matter interactions, which is becoming significant for fundamental quantum physics. In some scenarios, excitons in TMDs that cover visible and near-infrared wavelengths can couple with surface plasmons in metal to generate plexcitons, and when the light–matter coupling is strong enough, the Rabi oscillations and strong coupling regime would result in the formation of hybrid quasi-particles, called polaritons.

Strong coupling is made possible by the local optical density of photonic states (LDOS), which is primarily determined by the spatiotemporal confinement of electromagnetic fields, as characterized by the mode volume V and resonance Q-factor, LDOS α Q/V. Because of their direct band gap and large exciton binding energy, numerous systems integrating TMDs monolayer and plasmonic nanostructures have recently demonstrated robust plasmon–exciton coupling [6,7]. However, since many of the novelties and prospective applications of these monolayer TMDs are based on their excitonic light emission, having a controllable emission in such systems is essential for developing efficient photonic components. In addition, many optical applications require a significant amount of light absorption, and photonic crystals (PhCs) are one of the most robust platforms for boosting light absorption, modulating light emission, and improving light–matter interactions [8–10]. The integration of TMDs with PhC structure helps to maximize the light extraction by manipulating their excitonic emissions. PhC nanocavities have an exceptionally high Q factor of up to 10^6 and an ultrasmall mode volume (V) on the order of a cubic wavelength, allowing them to greatly increase the intensity of incident light. In general, resonant cavity can be formed in PhC by making point defects where light can be localized and trapped in the defect. The group velocity of light tends to become zero at the band edge of the PhC, resulting in the trapping of light at the band edge. When light interacts with a shock wave traveling through a one-dimensional PhC, the frequency shifts across the bandgap, narrowing the bandwidth and slowing light by orders of magnitude. This light trapping phenomenon can be used to couple band-edge resonance with other bound-state like excitons in order to achieve strong light–mater interactions. Strong coupling has been demonstrated between excitons in TMDs and optical bound states in the continuum (BICs) supported by PhC slab with a Rabi splitting of 5.4 meV [11]. The large oscillator strength of TMDs excitons, which leads to strong exciton–photon interactions and the formation of exciton polaritons, is responsible for the strong coupling in such systems. The existence of BIC in periodic PhC is triggered by the destructive interference of contrapropagating waveguide modes associated with the periodic potential. The BIC is characterized by the high-Q factor resulting in a drastic enhancement in the light–matter interaction phenomenon and is found to be beneficial in nonlinearly tunable devices [12].

In addition, the linear polarization of PhC might be coupled with the circular polarization (CP) of TMDs, resulting in a shift in the valley polarization of TMDs. Valleytronics in TMDs allows for the tuning of valley degrees of freedom, which opens up a lot of possibilities for encoding and manipulating data, leading to the realization of quantum devices and quantum computation [13]. However, because of the inter-valley scattering present in TMDs, the mechanism of valley depolarization is convoluted, resulting in a significant drop in the degree of valley polarization under room temperature [14]. Circular polarized light is very important for many important applications, such as circular dichroism spectroscopy [15], magnetic imaging [16], spintronics [17], quantum computing [18,19], optical communication [20], and manipulation of quantum states [21]. Controlling CP states of the light in TMDs will result in circular dichroism (CD), which is a phenomenon that occurs when light passes through a certain medium and splits into left-hand CP

(LCP) and right-hand CP (RCP) polarization states. Circular dichroism manipulation is significant in a variety of platforms, particularly in display technology, and CP radiation is becoming highly prevalent in chemistry, biology, and elementary physics apart from optics [22–26]. Due to the existence of CD, TMDs are particularly promising materials with controllable optical chirality for facilitating the generation of CP light. It has been challenging to manipulate the valley polarization in TMDs monolayer at room temperature due to weak light matter interaction and substantial defects. However, various approaches have been adopted to enhance the CD in these TMDs using in-plane electric field [27], out-of-plane magnetic field [2], localized magnetic field [28], plasmonic structures [29], etc. It has been reported that CD in TMDs materials, such as WSe_2, can be tuned by using plasmonic metasurfaces owing to the presence of localized surface plasmon resonance [28]. The enhanced light–matter interaction is responsible for the improvement of CD in these integrated systems. Photonic crystals can be integrated with TMDs to maximize the light extraction from TMDs through strong light–mater interactions. As a result, the integration between the monolayer TMDs and the PhC can control the polarization states of photons emitted from the TMDs. In our work, we use 1-D PhC, and its band-edge resonance is responsible for coupling with the excitons in TMDs monolayer thereby enhancing the light–mater interaction in the integrated system.

Tungsten diselenide (WSe_2) and molybdenum disulfide (MoS_2) are two TMDs materials that have been the most investigated in recent years [30–34]. However, MoS_2 has the downside of being easily oxidized in air, resulting in S vacancies in the material, thereby lowering its optical quality [35]. WSe_2, on the other hand, drew a lot of attention because to its distinctive features, such as high quantum yield, strong spin-orbit coupling, and ambipolar charge transport. In addition, the peak emission wavelength, band gap and PL intensity of WSe_2 change with the change in the number of layers. In this paper, we show how to tune the optical mode of a 1-D PhC on a flexible PDMS substrate by increasing the PhC lattice constant. Furthermore, we use a monolayer layer of WSe_2 with a thickness of only 0.7 nm as the gain material, which can couple with the resonant mode of the 1-D PhC cavity. The magnitude of degree of polarization is significantly enhanced through the exciton–photon interaction with the integration of 1-D PhC. This work could pave a way towards the manipulation of a degree of circular polarization in various TMDs integrated systems.

2. Materials and Methods

2.1. Fabrication of SiN_x 1-D Photonic Crystals

Low pressure chemical vapor deposition (LPCVD) was used to deposit SiN_x with a thickness of 200 nm on a silicon substrate, followed by spin coating of 300 nm thick ma-N2403 photoresist and a thin layer of espacer on our SiN_x/Si substrate. Due to its better scale and resolution to our pattern, electron beam lithography was a good candidate for our device fabrication and was used to pattern our SiN_x/Si substrate. For the e-beam process, the spot size and the beam voltage were set at 1.0 and 30 kV when patterning, and the exposed ma-N2403 was developed by the tetramethylammonium hydroxide (TMAH 2.38%) solution. We used high-density plasma inductively coupled plasma-reactive ion etching (ICP-RIE) dry etching system for transferring the pattern into the SiN_x PhC layer. After the etching was completed, the residual ma-N2403 photoresist was removed from the substrate through O_2 plasma in ICP-RIE at 20 °C with O_2 flow of 35 sccm for 5 min to finally obtain the SiN_x 1-D PhC. Figure 1a illustrates the fabrication process of the SiN_x PhC and its SEM image and optical microscopy image are shown in Figure 1b.

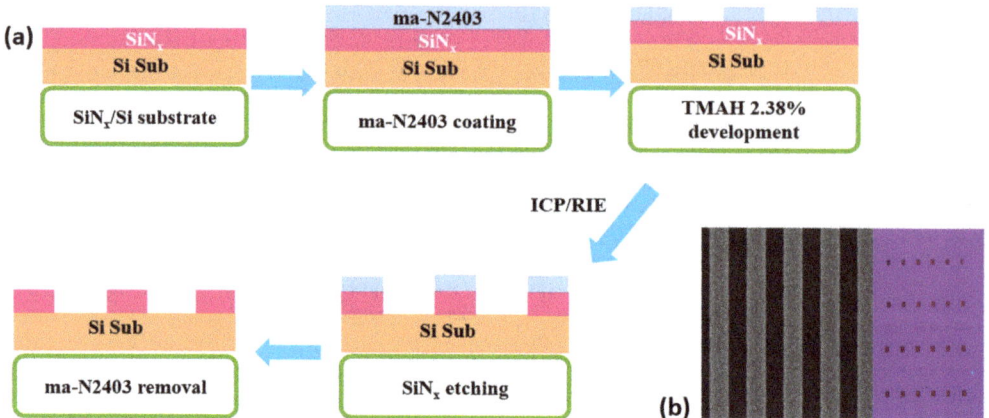

Figure 1. (a) SiN$_x$ 1-D photonic crystal fabrication steps. For transferring the pattern, ICP/RIE etching technology was used. (b) SEM image and optical microscope image of the SiN$_x$ 1-D PhC.

2.2. Transfer Process of 1-D PhC Structure on Flexible Substrate

We employed the most common PDMS elastomers, Sylgard® 184 from Dow Corning® (Midland, MI, USA), which has two resin components with vinyl groups (Part A) and hydrosiloxane groups (Part B), for the preparation of the flexible substrate. The PDMS elastomer was cured by mixing the two resin components A and B solutions with a volume ratio of 10:1 followed by the elimination of bubbles from PDMS in vacuum environment for 1 h. Then, we poured 2 mL of the PDMS solution into a plastic Petri dish and heated it at 75 °C for 12 min to keep the sample from sinking to the bottom and the substrate was finally half-cured. We then added 1 mL of uncured PMDS solution to finish the PDMS substrate, which was subsequently bonded to an upside down SiN$_x$/Si structure and heated for 30 min at 75 °C. After the PDMS substrate was constructed, the Si substrate was removed using a diluted TMAH solution (2.38%), followed by a DI water rinse and hot plate drying to obtain the SiN$_x$ pattern on the PDMS substrate. The process flow of bonding the SiN$_x$ 1-D PhC structure on the PDMS substrate is demonstrated in Figure 2a and the optical microscopy image of the patterned PDMS substrate with SiN$_x$ is shown in Figure 2b.

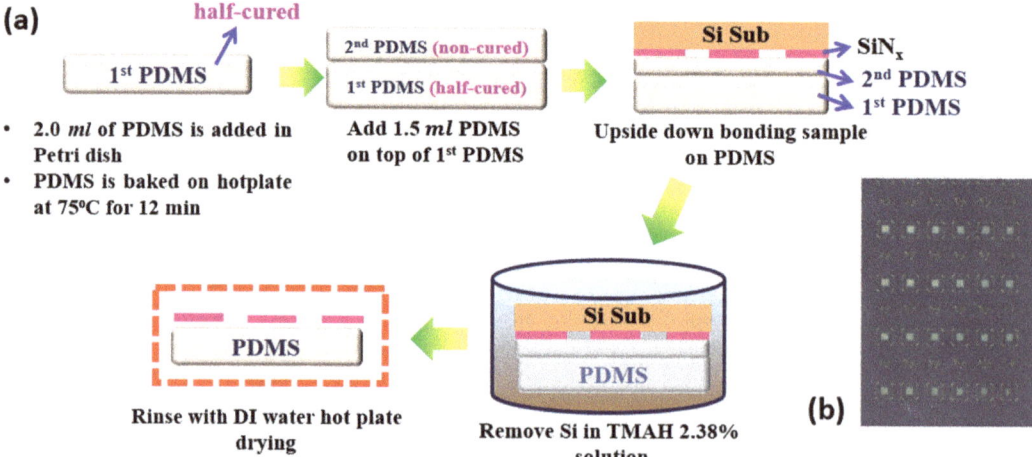

Figure 2. (a) Process flow chart of bonding the SiN$_x$ pattern on PDMS. (b) Optical image of the fabricated structure on the PDMS substrate.

2.3. Transfer Process of WSe$_2$ Monolayer

To transfer WSe$_2$ to the flexible substrate, the monolayer WSe$_2$ was first grown on a sapphire substrate by chemical vapor deposition (CVD). The process flow of the transfer process of WSe$_2$ to the flexible substrate is shown in Figure 3. Before transferring, we spin-coated a layer of PMMA A5 on the top of WSe$_2$ monolayer with the sapphire substrate at 1000 RPM for 1 min and baked at 100 °C for 30 min. The substrate was then immersed for 90 min in hot buffered oxide etch (BOE) at 100 °C to etch the sapphire and establish a narrow gap between the PMMA/WSe$_2$ layer and sapphire substrate. The substrate was immersed in DI water after the etching procedure to remove BOE, and the PMMA/WSe$_2$ layer was easily separated from the sapphire substrate and floated on the DI water surface. The PMMA/WSe$_2$ layer was then picked up using our flexible substrate, and we ensured that the WSe$_2$ flakes were overlapped with the 1-D PhC structures. The sample was then obliquely baked at 100 °C for 12 h before being immersed in acetone for 15 min to remove the PMMA coating. Finally, we baked the sample at 100 °C to dry the substrate, and an optical image of the WSe$_2$ monolayer/1-D PhC structure on a flexible substrate is shown at the end of the last transfer step.

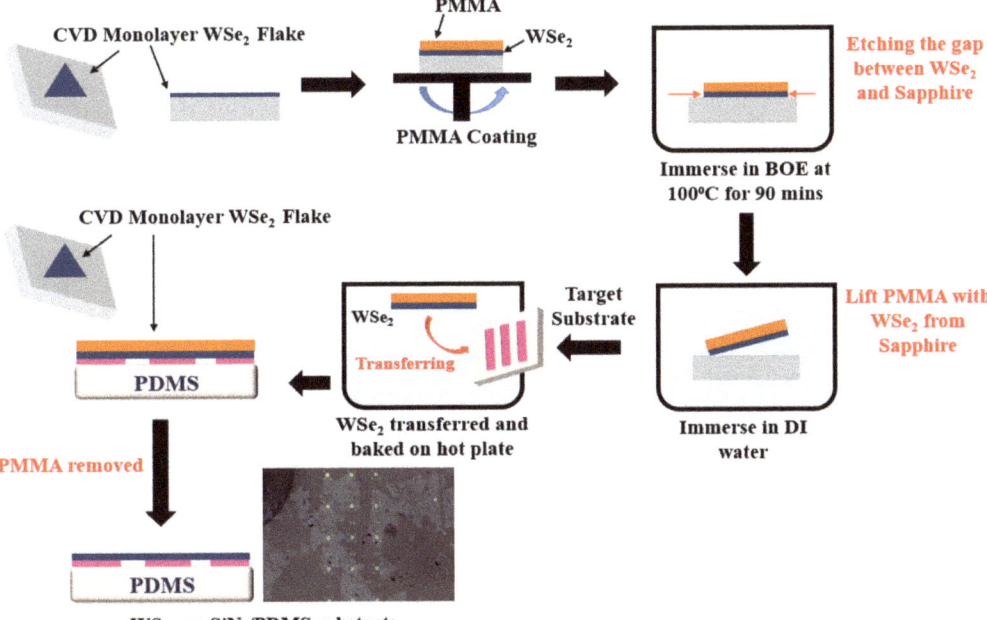

Figure 3. Schematic of the transfer process flow of WSe$_2$ to the flexible substrate through the PMMA-assisted transfer. An optical image following the PMMA transfer for the WSe$_2$ monolayer on SiN$_x$/PDMS substrate is shown at the end of the last transfer step.

3. Results and Discussion

For many years, photonic crystals have attracted attention, and in a one-dimensional photonic bandgap, the light at the band edge has practically zero group velocity, allowing for lasing. A PhC structure with uniform periodicity can be used to couple the light with the gain material. In this work, we used a 1-D SiN$_x$ PhC structure having different lattice constant that varies from 410 nm to 470 nm, and the filling factor in each lattice constant is changed. The band-edge resonance of 1-D PhC couples with the excitons of WSe$_2$ monolayer with an emission wavelength of 750 nm and a bandgap of 1.65 eV. For the PhC, the wavelength corresponds to the parameters of the lattice constant as shown in Figure 4a and the wavelength increases with the period in the cavity of the PhC structure. The resonant cavity mode of the PhC structure couples with WSe$_2$ excitons once it is transferred to the flexible substrate, as illustrated in Figure 4b, and the PhC peaks are similar in both cases. We measured a series of devices with different periods to verify the band-edge resonant modes in the 1-D PhCs, and we expected the normalized frequency (a/λ) to be the same in the same filling factor for a band-edge mode. As shown in Figure 4c, the wavelength increases linearly with the lattice constant and the normalized frequency is approximately 0.61, indicating that the mode is the same. In order to investigate the optical modes of the 1-D PhC on a flexible PDMS substrate, we calculated the corresponding band structure of the PhC by the plane-wave expansion (PWE) method for TE-like modes. Figure 4d shows the simulated band structure with a period of 470 nm and a width of 269 nm showing the group velocity of light with different in-plane wave vector k from along the x-axis to along the y-axis. In fact, the PhC band-edge mode is likely to occur around high-symmetry points, and the flat photonic band with low group velocity can enhance the light interaction. By comparing the experimental and simulated results, we can be certain that the circle in the band structure represents the operation mode of the PhC, which has a normalized frequency of 0.61.

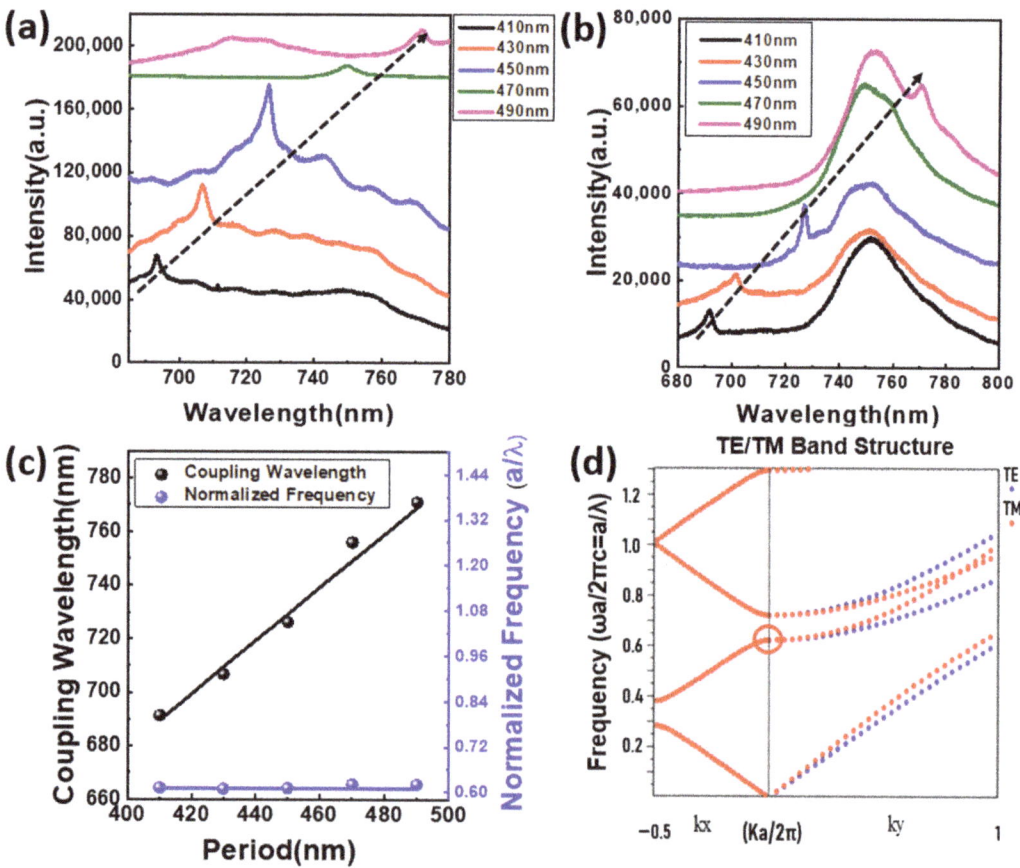

Figure 4. Coupling wavelength of the (**a**) 1-D PhC with different periods. (**b**) Integrated system with different lattice periods. (**c**) Coupling wavelength and normalized frequency with different lattice periods. (**d**) The 2-D PWE simulated TE-like band structure of the PhC.

We tried to investigate the characteristics of the integrated system with various filling factors after ensuring that the PhC operating mode is a band-edge resonant mode. We used SEM to determine the size of the PhC structure (4–1 to 4–6) as shown in Figure 5a, with filling factors ranging from 0.46 to 0.58. Figure 5b depicts the spectrum of devices with varying filling factors, but the same lattice constant of 470 nm and height of 200 nm, implying that the wavelength shifts depending on the filling factor ratio. It can be seen that the cavity mode and gain material of WSe$_2$ have a high anti-crossing dispersion relationship, which is an essential Rabi splitting phenomenon. When the emitter–photon interaction becomes larger than the dissipation rates of the system, this enables quantum coherent oscillations between the coupled systems and the quantum superposition between different quantum states. The Rabi splitting of the coupled system is around 25.0 meV, indicating that the band-edge resonance of 1-D PhC and the excitons in WSe$_2$ monolayer were coupled successfully. We can notice that the coupling wavelength increases as the filling factor increases because the corresponding normalized frequency increases as shown in Figure 5c. In the 1-D PhC structure, the effective index increases as the filling factor increases, causing the normalized frequency to decrease and coupling wavelength to be red-shifted.

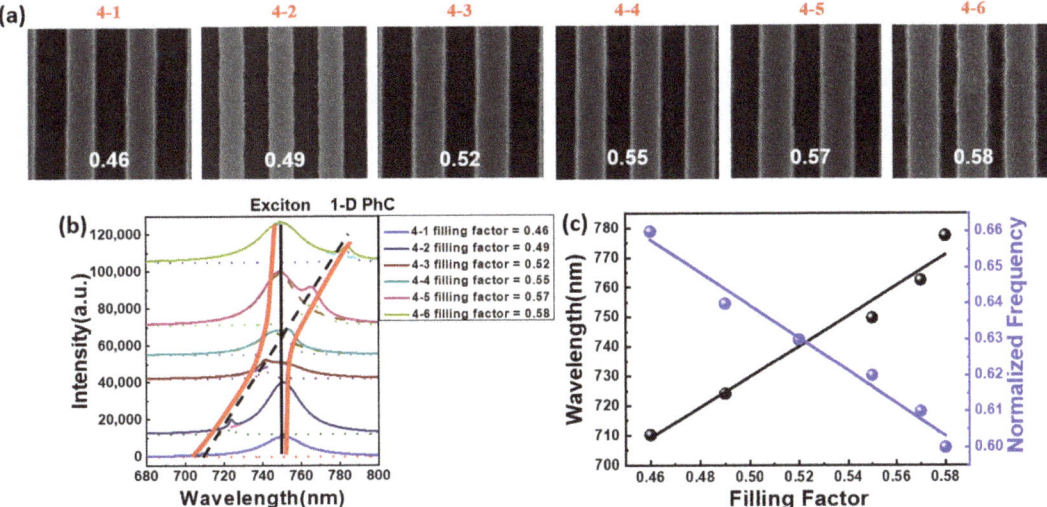

Figure 5. (a) Different 1-D PhCs (4–1 to 4–6) with various filling factors. (b) Spectrum of the coupled system for different filling factor ratio in the same period of 470 nm. (c) Coupling wavelength and corresponding normalized frequency for different filling factors.

To manipulate the optical characteristics of the WSe$_2$ integrating with the flexible SiN$_x$ 1-D PhC structure, we used a homemade extending stage for stretching the flexible substrate as shown in Figure 6a. The sample is fixed on the stage by two clamps, and the sample can be stretched in the lateral direction by rotating the micrometer. By using an optical microscope and comparing the variation of length to the original length, we were able to measure the lattice extension in terms of percentage. When the PhC structure is stretched, the strain on SiN$_x$ PhC structure and PDMS substrate can result in numerous changes. Young's modulus for SiN$_x$ and PDMS are 297 GPa and 870 KPa, respectively, hence the SiN$_x$ PhC structure has less deformation than PDMS. In addition, the Poisson's ratio while the strain is applied as shown in the inset of Figure 6a can be calculated as

$$\vartheta = -\frac{d\varepsilon_{trans}}{d\varepsilon_{axial}} \qquad (1)$$

where $d\varepsilon_{trans}$ is the axial strain while changing the length of x-direction, positive when stretching; and $d\varepsilon_{axial}$ is the short-axis strain while changing the length of y-direction, negative when stretching. Figure 6b shows that the Poisson's ratio is approximately 0.55 after a series of measurements.

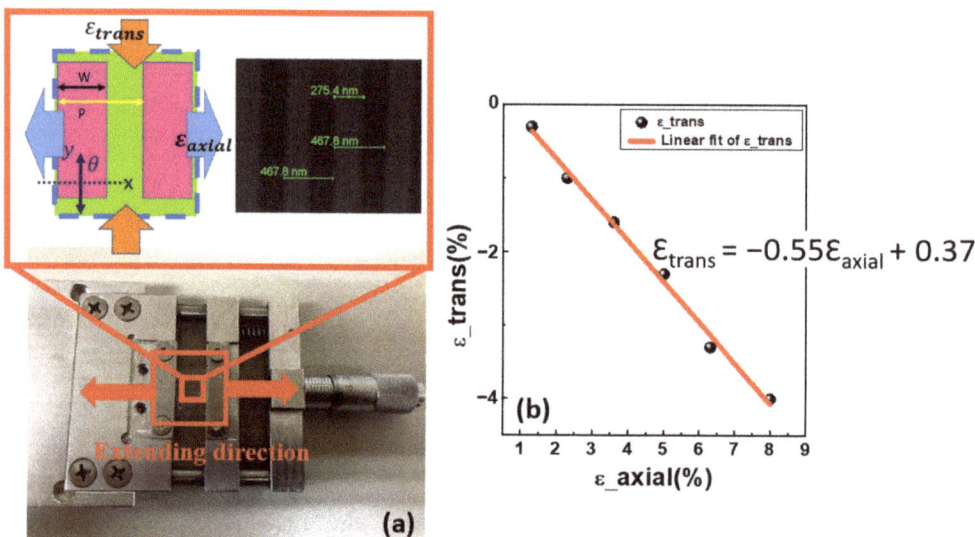

Figure 6. (a) Homemade stage that serves as the extending platform. (b) Poisson's ratio calculation.

The pattern's lattice constant in this experiment is 470 nm, which increases as the pattern stretches. The extension of the lattice constant along the x-direction is measured by optical microscope and the strain percentage is defined as $X (\%) = (X - X_0)/X_0 \times 100\%$, where X_0 and X are the length of the original pattern bonded on the PDMS substrate and after deformation, respectively. In this experiment, the strain percentage reflects the relative deformation in the size of a single pattern, which is 30×30 μm^2. The period increases from 470 nm to 506 nm as the pattern's strain increases to 8%, as seen in Table 1. In addition, as the strain of the pattern decreases to 1.6%, the period decreases from 470 nm to 462 nm. The pattern's period is linearly proportional to the pattern's relative deformation, according to the measurements.

Table 1. Estimating the period of the 1-D PhC with different strains from the optical microscope.

Strain	−1.6%	0%	1.3%	2.3%	3.6%	5.0%	6.3%	8.0%
Period	460 nm	470 nm	475 nm	479 nm	485 nm	492 nm	498 nm	506 nm

We created the structure on a flexible PDMS substrate because we wanted to expand it so that the geometry of the structure could be fine-tuned, and the wavelength can change as the period is extended in the flexible substrate. Figure 7a represents the PL spectra of the coupled system with different strains as the system is stretched in the x-direction. As we can see, the peak of the PhC structure can be manipulated as the strain is increased or decreased. From Figure 7b, we can see that, with the strain variation from −1.6% to 6.3%, the wavelength can be fine-tuned and increases linearly with the red-shifted wavelength, which is attributed to the period extension of the PhC structure on the flexible substrate. However, the PL peak wavelength of WSe$_2$ does not shift with the increase in the strain as shown in Figure 7c. In addition, in the Raman spectra of Figure 7d, the peak positions of E^1_{2g} and A_{1g} do not split and remain constant as the PDMS strain increases. The main reason is due to a strain transfer problem caused by a large variation in the Young's modulus of the PDMS (870 KPa) substrate and the monolayer WSe$_2$ (258.6 GPa), resulting in a lower strain transfer efficiency. As a result, the WSe$_2$ monolayer is harder than the PDMS substrate and hence the monolayer cannot be deformed by the flexible substrate. This problem can be overcome by using a flexible substrate with a Young's modulus comparable to that of the monolayer WSe$_2$, which improves the strain transfer efficiency.

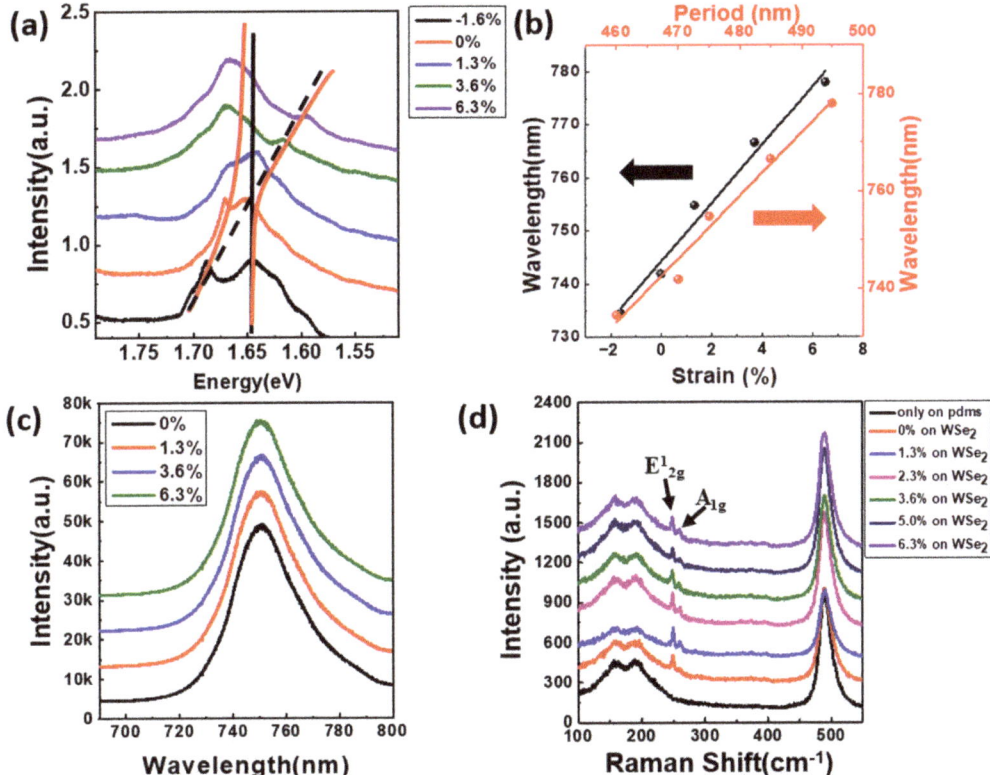

Figure 7. (**a**) PL spectrum of the PhC structure integrated with WSe$_2$ for various strains. (**b**) Wavelength shift with strain and period. (**c**) PL spectra of WSe$_2$ with different strains. (**d**) Raman spectra of WSe$_2$ with different strains.

The existence of circular dichroism in TMDs enables them to control their optical chirality; there is a significant interest and potential applications for chiral 2D nanomaterials. There are many approaches to maintain the CD of PL in TMDs, such as applications of electric field and magnetic fields [2,27,36]. In this paper, we show the manipulation of CD of WSe$_2$ through integration with 1-D PhC structure. Since the PhC can boost light absorption and improve the light–matter interactions by manipulating the excitonic emission, it can also be utilized to manipulate the degree of circular polarization in TMDs. In general, 1-D PhC exhibit band-edge resonance where the density of states at the resonant frequency is very high. This can be utilized to couple with excitons in the monolayer WSe$_2$ for an enhanced light–mater interaction, which in turn can improve the degree of circular polarization for WSe$_2$. In general, the applications of 2D materials in valleytronics have become very significant, and the circular polarization state can be applied to encode data for optical communications. WSe$_2$ is an appropriate candidate for inducing optical chirality with the integration of nanostructures. The band structures of TMDs consist of two inequivalent valleys, i.e., K and −K, which lead to strong spin–orbit coupling. As a result, according to the valley-dependent optical selection rule, right circular polarized light can couple to excitonic transitions in the K valley, whereas left circular polarized light can couple to excitonic transitions in the –K valley, resulting in the single handedness of the corresponding light emission. However, as shown in Figure 8, the PL emission from the WSe$_2$ after RCP light excitation contains not only RCP light from K valley, but also LCP light corresponding to the –K valley. This is attributed to the inter-valley scattering of excitons between the K and –K valley originated from optical phonons induced by defects. This is the reason why WSe$_2$ had a smaller degree of polarization at room temperature. However, when WSe$_2$ is integrated with a 1-D PhC structure, the optical absorption is enhanced resulting in the generation of more excitons under RCP light excitation in the K valley. These excitons lead to the improvement in the optical activities that contribute to the enhancement in the degree of valley polarization. In addition, the inter-valley scattering between the K and –K valleys is reduced due to strong light–matter interactions, thereby suppressing the excitons generation in –K valley.

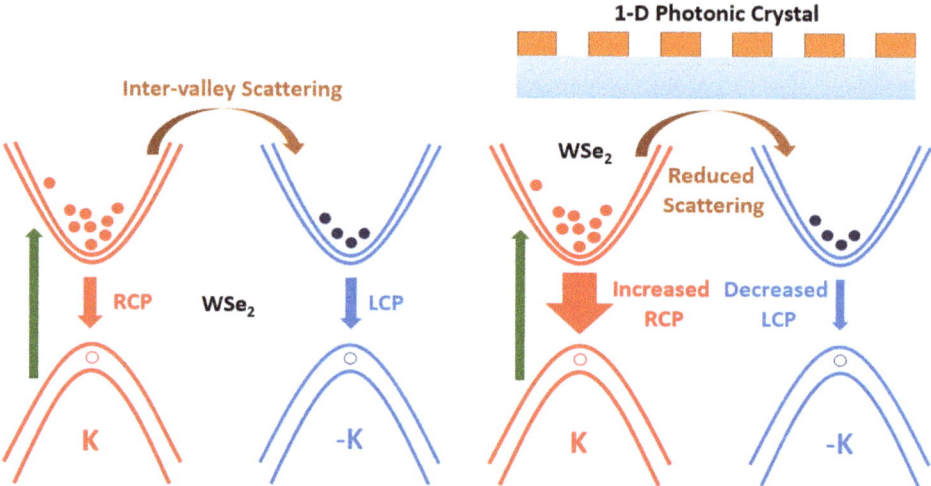

Figure 8. Mechanism for monitoring valley polarization in WSe$_2$/1-D PhC integrated system. Under RCP light excitation, the WSe$_2$/1-D PhC integrated system shows a significantly higher decay rate for excitons in K valley, while reduced decay rates for excitons in –K valley.

The degree of valley polarization can be defined using the following formula:

$$P = \frac{I_R - I_L}{I_R + I_L} \times 100\% \tag{2}$$

where I_R and I_L represent the intensities of right- and left-polarized light, respectively. The degree of valley polarization depends on the excitons population with their decay rates and the inter-valley scattering between the two valleys. As a result, the degree of valley polarization can be enhanced by increasing the excitons populations or decay rates and suppressing the inter-valley scattering. The presence of 1-D PhC leads to strong light–matter interactions due to the coupling between excitons and the band-edge resonance, which in turn increases the excitons' decay rates and reduces the inter-valley scattering, thereby improving the degree of valley polarization in the WSe$_2$ monolayer/1-D PhC integrated system.

In the experiment, we characterized the circular polarization of the exciton absorption from the bare WSe$_2$ and the WSe$_2$/1-D PhC devices. Figure 9a show the absorption spectra of the bare WSe$_2$ exciton under the LCP light pumping. The exciton absorption of the bare WSe$_2$ prefers left-hand polarization due to the valleytronic effect in the WSe$_2$ monolayer. Figure 9b shows the absorption spectra of the exciton from the WSe$_2$/1-D PhC device under the LCP light pumping. The difference between the LCP and RCP absorption of the WSe$_2$ exciton was much more enhanced after the integration of 1-D PhC. Figure 9c shows the circular polarization degree (CD) of the exciton absorption estimated with Equation (2), and a CD value of approximately −2.44% was observed for the bare WSe$_2$ monolayer under LCP light pumping. The CD spectrum of the WSe$_2$/1-D PhC is shown in Figure 9d, and an exciton absorption with a CD value of approximately −20.44% was achieved under LCP light pumping conditions, while the LCP light absorption was higher than that of RCP light absorption for the WSe$_2$/1-D PhC structure under LCP pumping due to the introduction of the band-edge resonant mode from 1-D PhC. The difference in the absorption spectra demonstrates the photoinduced nonreciprocal dichroic behavior in the WSe$_2$/1-D PhC system owing to the coupling between excitons in WSe$_2$ and band-edge resonance of the 1-D PhC. The mechanism related to this nonreciprocity is associated with the difference in the density of the excitons in the two valleys of WSe$_2$ when coupled with the resonances of 1-D PhC. The degree of valley polarization increased from −2.44% to −20.44% under LCP light incidence. A similar phenomenon with the inverse CDs was also observed for the bare WSe$_2$ and the WSe$_2$/1-D PhC devices under the RCP pumping. The significantly high degree of valley polarization values from the WSe$_2$/1-D PhC integrated system is attributed to the light–matter interaction in the coupled system resulting from the coupling between excitons and the PhC resonant mode. Hence, the optical chirality of the WSe$_2$ monolayer emission can be manipulated by integrating the 1-D PhC structure with the WSe$_2$ monolayer.

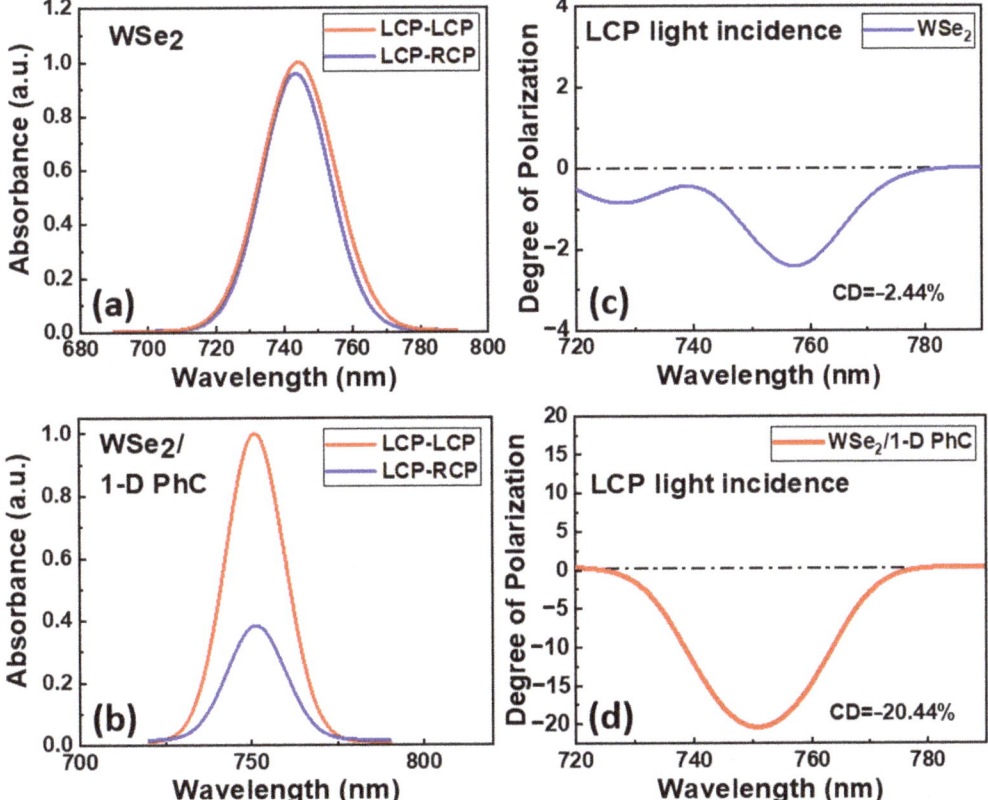

Figure 9. (**a**,**b**) Circularly polarized light absorption spectra for bare WSe$_2$ and WSe$_2$/1-D PhC under LCP light incidence. (**c**,**d**) Degree of polarization for WSe$_2$ and WSe$_2$/1-D PhC under LCP light incidence, respectively.

4. Conclusions

In summary, we investigated the coupling between the excitonic emission of a WSe2 monolayer and the band-edge resonance of 1-D PhC on a PDMS flexible substrate. The Rabi splitting of the coupled system was around 25.0 meV, indicating that coupling exists between the 1-PhC resonance and WSe$_2$ excitons in the integrated system. Furthermore, we demonstrated the manipulation of the circular polarized light emission through the light–matter interaction between the WSe$_2$ excitons and the PhC band-edge resonance. In particular, we demonstrated the enhancement of the degree of circular polarization from −2.44% to −20.44% under LCP light incidence after the integration of 1-D PhC. A similar behavior with inverse CD is also observed under RCP light pumping. The stronger light–matter interaction in the coupled system is shown to control the valley-dependent excitons generation thereby manipulating the valley polarization in the WSe$_2$ monolayer. A device capable of efficiently modulating circular polarized photon emission was developed by combining a novel gain material with low-dimensional materials, such as the 1-D PhC structure. These results also provide a possible platform to control the circular polarized states of the chip-scale emitters for various applications, such as optical information technology, biosensing and various opto-valleytronic devices based on 2D materials.

Author Contributions: Conceptualization, K.J.S., H.-H.C., Y.-H.C., Y.-S.L., H.-T.L., M.-H.S. and H.-C.K.; data curation, K.J.S., H.-H.C., Y.-H.C., Y.-S.L., H.-T.L., M.-H.S. and H.-C.K.; formal analysis, K.J.S., H.-H.C., M.-H.S. and H.-C.K.; investigation, K.J.S., H.-H.C., M.-H.S. and H.-C.K.; Methodology, K.J.S., H.-H.C., Y.-H.C., Y.-S.L., P.-C.T. and S.-Y.L.; project administration, M.-H.S. and H.-C.K.; resources, P.-C.T. and S.-Y.L.; supervision, M.-H.S. and H.-C.K.; validation, H.-T.L., M.-H.S. and H.-C.K.; visualization, M.-H.S. and H.-C.K.; writing—original draft, K.J.S.; writing—review and editing, K.J.S., M.-H.S. and H.-C.K. All authors have read and agreed to the published version of the manuscript.

Funding: This research was funded by Ministry of Science and Technology, Taiwan (MOST 108-2112-M-001-044-MY2 and MOST 110-2112-M-001-053-).

Institutional Review Board Statement: Not applicable.

Informed Consent Statement: Not applicable.

Data Availability Statement: Not applicable.

Acknowledgments: This work was supported by the Innovative Materials and Analytical Technology Exploration (i-MATE) program of Academia Sinica in Taiwan and the Ministry of Science and Technology (MOST) in Taiwan under Contract number MOST 108-2112-M-001-044-MY2 and MOST 110-2112-M-001-053-.

Conflicts of Interest: The authors declare no conflict of interest.

References

1. Jiang, Y.; Chen, S.; Zheng, W.; Zheng, B.; Pan, A. Interlayer exciton formation, relaxation, and transport in TMD van der Waals heterostructures. *Light Sci. Appl.* **2021**, *1*, 72. [CrossRef] [PubMed]
2. Mak, K.F.; Shan, J. Photonics and optoelectronics of 2D semiconductor transition metal dichalcogenides. *Nat. Photonics* **2016**, *10*, 216–226. [CrossRef]
3. Mak, K.F.; He, K.; Lee, C.; Lee, G.-H.; Hone, J.; Heinz, T.F.; Shan, J. Tightly bound trions in monolayer MoS$_2$. *Nat. Mater.* **2012**, *12*, 207–211. [CrossRef]
4. Xiao, D.; Liu, G.-B.; Feng, W.; Xu, X.; Yao, W. Coupled Spin and Valley Physics in Monolayers of MoS$_2$ and Other Group-VI Dichalcogenides. *Phys. Rev. Lett.* **2012**, *108*, 196802. [CrossRef]
5. Kuc, A.; Zibouche, N.; Heine, T. Influence of quantum confinement on the electronic structure of the transition metal sulfide TS2. *Phys. Rev. B* **2011**, *83*, 245213. [CrossRef]
6. Fandan, R.; Pedrós, J.; Calle, F. Exciton-Plasmon Coupling in 2D Semiconductors Accessed by Surface Acoustic Waves. *ACS Photonics* **2021**, *8*, 1698–1704. [CrossRef]
7. Liu, L.; Tobing, L.Y.M.; Yu, X.; Tong, J.; Qiang, B.; Fernández-Domínguez, A.I.; Garcia-Vidal, F.J.; Zhang, D.H.; Wang, Q.J.; Luo, Y. Strong Plasmon–Exciton Interactions on Nanoantenna Array-Monolayer WS$_2$ Hybrid System. *Adv. Opt. Mater.* **2019**, *8*, 1901002. [CrossRef]
8. David, A.; Benisty, H.; Weisbuch, C. Photonic crystal light-emitting sources. *Rep. Prog. Phys.* **2012**, *75*, 126501. [CrossRef]
9. Gan, X.; Mak, K.F.; Gao, Y.; You, Y.; Hatami, F.; Hone, J.; Heinz, T.F.; Englund, D. Strong Enhancement of Light–Matter Interaction in Graphene Coupled to a Photonic Crystal Nanocavity. *Nano Lett.* **2012**, *12*, 5626–5631. [CrossRef] [PubMed]

10. Majumdar, A.; Kim, J.; Vuckovic, J.; Wang, F. Electrical Control of Silicon Photonic Crystal Cavity by Graphene. *Nano Lett.* **2013**, *13*, 515–518. [CrossRef]
11. Koshelev, K.L.; Sychev, S.K.; Sadrieva, Z.F.; Bogdanov, A.A.; Iorsh, I.V. Strong coupling between excitons in transition metal dichalcogenides and optical bound states in the continuum. *Phys. Rev. B* **2018**, *98*, 161113. [CrossRef]
12. Krasikov, S.D.; Bogdanov, A.A.; Iorsh, I.V. Nonlinear bound states in the continuum of a one-dimensional photonic crystal slab. *Phys. Rev. B* **2018**, *97*, 224309. [CrossRef]
13. Liu, Y.; Gao, Y.; Zhang, S.; He, J.; Yu, J.; Liu, Z. Valleytronics in transition metal dichalcogenides materials. *Nano Res.* **2019**, *12*, 2695–2711. [CrossRef]
14. Zeng, H.; Dai, J.; Yao, W.; Xiao, D.; Cui, X. Valley polarization in MoS_2 monolayers by optical pumping. *Nat. Nanotechnol.* **2012**, *7*, 490–493. [CrossRef] [PubMed]
15. Ranjbar, B.; Gill, P. Circular Dichroism Techniques: Biomolecular and Nanostructural Analyses—A Review. *Chem. Biol. Drug Des.* **2009**, *74*, 101–120. [CrossRef]
16. Kfir, O.; Zayko, S.; Nolte, C.; Sivis, M.; Möller, M.; Hebler, B.; Arekapudi, S.S.P.K.; Steil, D.; Schäfer, S.; Albrecht, M.; et al. Nanoscale magnetic imaging using circularly polarized high-harmonic radiation. *Sci. Adv.* **2017**, *3*, eaao464. [CrossRef]
17. Miyamoto, K.; Wortelen, H.; Okuda, T.; Henk, J.; Donath, M. Circular-polarized-light-induced spin polarization characterized for the Dirac-cone surface state at W(110) with C2v symmetry. *Sci. Rep.* **2018**, *8*, 10440. [CrossRef]
18. Togan, E.; Chu, Y.; Trifonov, A.S.; Jiang, L.; Maze, J.; Childress, L.; Dutt, M.V.G.; Sørensen, A.S.; Hemmer, P.R.; Zibrov, A.S.; et al. Quantum entanglement between an optical photon and a solid-state spin qubit. *Nature* **2010**, *466*, 730–734. [CrossRef]
19. Wagenknecht, C.; Li, C.-M.; Reingruber, A.; Bao, X.-H.; Goebel, A.; Chen, Y.-A.; Zhang, Q.; Chen, K.; Pan, J.-W. Experimental demonstration of a heralded entanglement source. *Nat. Photonics* **2010**, *4*, 549–552. [CrossRef]
20. Farshchi, R.; Ramsteiner, M.; Herfort, J.; Tahraoui, A.; Grahn, H.T. Optical communication of spin information between light emitting diodes. *Appl. Phys. Lett.* **2011**, *98*, 162508. [CrossRef]
21. Fujita, T.; Morimoto, K.; Kiyama, H.; Allison, G.; Larsson, M.; Ludwig, A.; Valentin, S.R.; Wieck, A.D.; Oiwa, A.; Tarucha, S. Angular momentum transfer from photon polarization to an electron spin in a gate-defined quantum dot. *Nat. Commun.* **2019**, *10*, 2991. [CrossRef] [PubMed]
22. Agranat, I.; Caner, H.; Caldwell, J. Putting chirality to work: The strategy of chiral switches. *Nat. Rev. Drug Discov.* **2002**, *1*, 753–768. [CrossRef] [PubMed]
23. Albani, L.; Marchessoux, C.; Kimpe, T. Stereoscopic Display Technologies and Their Applications in Medical Imaging. *Inf. Disp.* **2011**, *27*, 24–29. [CrossRef]
24. Kelly, S.M.; Jess, T.J.; Price, N.C. How to study proteins by circular dichroism. *Biochim. Biophys. Acta (BBA)—Proteins Proteom.* **2005**, *1751*, 119–139. [CrossRef] [PubMed]
25. Monti, S.; Manet, I.; Marconi, G. Combination of spectroscopic and computational methods to get an understanding of supramolecular chemistry of drugs: From simple host systems to biomolecules. *Phys. Chem. Chem. Phys.* **2011**, *13*, 20893–20905. [CrossRef]
26. Whitmore, L.; Wallace, B.A. Protein secondary structure analyses from circular dichroism spectroscopy: Methods and reference databases. *Biopolymers* **2008**, *89*, 392–400. [CrossRef]
27. Wu, S.; Ross, J.S.; Liu, G.-B.; Aivazian, G.; Jones, A.; Fei, Z.; Zhu, W.; Xiao, D.; Yao, W.; Cobden, D.; et al. Electrical tuning of valley magnetic moment through symmetry control in bilayer MoS2. *Nat. Phys.* **2013**, *9*, 149–153. [CrossRef]
28. Lin, H.-T.; Chang, C.-Y.; Cheng, P.J.; Li, M.-Y.; Cheng, C.-C.; Chang, S.-W.; Li, L.L.J.; Chu, C.W.; Wei, P.-K.; Shih, M.-H. Circular Dichroism Control of Tungsten Diselenide (WSe_2) Atomic Layers with Plasmonic Metamolecules. *ACS Appl. Mater. Interfaces* **2018**, *10*, 15996–16004. [CrossRef]
29. Lin, W.-H.; Wu, P.C.; Akbari, H.; Rossman, G.R.; Yeh, N.-C.; Atwater, H.A. Electrically Tunable and Dramatically Enhanced Valley-Polarized Emission of Monolayer WS_2 at Room Temperature with Plasmonic Archimedes Spiral Nanostructures. *Adv. Mater.* **2021**, *34*, 2104863. [CrossRef]
30. As'ham, K.; Al-Ani, I.; Huang, L.; Miroshnichenko, A.E.; Hattori, H.T. Boosting Strong Coupling in a Hybrid WSe_2 Monolayer–Anapole-Plasmon System. *ACS Photonics* **2021**, *8*, 489–496. [CrossRef]
31. Cao, L.; Zhong, J.; Yu, J.; Zeng, C.; Ding, J.; Cong, C.; Yue, X.; Liu, Z.; Liu, Y. Valley-polarized local excitons in WSe_2/WS_2 vertical heterostructures. *Opt. Express* **2020**, *28*, 22135–22143. [CrossRef] [PubMed]
32. Datta, K.; Li, Z.; Lyu, Z.; Deotare, P.B. Piezoelectric Modulation of Excitonic Properties in Monolayer WSe_2 under Strong Dielectric Screening. *ACS Nano* **2021**, *15*, 12334–12341. [CrossRef] [PubMed]
33. Minn, K.; Anopchenko, A.; Chang, C.-W.; Mishra, R.; Kim, J.; Zhang, Z.; Lu, Y.-J.; Gwo, S.; Lee, H.W.H. Enhanced Spontaneous Emission of Monolayer MoS_2 on Epitaxially Grown Titanium Nitride Epsilon-Near-Zero Thin Films. *Nano Lett.* **2021**, *21*, 4928–4936. [CrossRef] [PubMed]
34. Qin, C.; Gao, Y.; Zhang, L.; Liang, X.; He, W.; Zhang, G.; Chen, R.; Hu, J.; Xiao, L.; Jia, S. Flexible engineering of light emission in monolayer MoS_2 via direct laser writing for multimode optical recording. *AIP Adv.* **2020**, *10*, 045230. [CrossRef]
35. Kc, S.; Longo, R.C.; Wallace, R.M.; Cho, K. Surface oxidation energetics and kinetics on MoS_2 monolayer. *J. Appl. Phys.* **2015**, *117*, 135301. [CrossRef]
36. Ye, Y.; Xiao, J.; Wang, H.; Ye, Z.; Zhu, H.; Zhao, M.; Wang, Y.; Zhao, J.; Yin, X.; Zhang, X. Electrical generation and control of the valley carriers in a monolayer transition metal dichalcogenide. *Nat. Nanotechnol.* **2016**, *11*, 598–602. [CrossRef]

Communication

Atomic Arrangements of Graphene-like ZnO

Jong Chan Yoon [1,2]**, Zonghoon Lee** [1,2] **and Gyeong Hee Ryu** [3,*]

[1] Department of Materials Science and Engineering, Ulsan National Institute of Science and Technology (UNIST), Ulsan 44919, Korea; yjc1526@unist.ac.kr (J.C.Y.); zhlee@unist.ac.kr (Z.L.)
[2] Center for Multidimensional Carbon Materials, Institute for Basic Science (IBS), Ulsan 44919, Korea
[3] School of Materials Science and Engineering, Gyeongsang National University, Jinju 52828, Korea
* Correspondence: gh.ryu@gnu.ac.kr

Abstract: ZnO, which can exist in various dimensions such as bulk, thin films, nanorods, and quantum dots, has interesting physical properties depending on its dimensional structures. When a typical bulk wurtzite ZnO structure is thinned to an atomic level, it is converted into a hexagonal ZnO layer such as layered graphene. In this study, we report the atomic arrangement and structural merging behavior of graphene-like ZnO nanosheets transferred onto a monolayer graphene using aberration-corrected TEM. In the region to which an electron beam is continuously irradiated, it is confirmed that there is a directional tendency, which is that small-patched ZnO flakes are not only merging but also forming atomic migration of Zn and O atoms. This study suggests atomic alignments and rearrangements of the graphene-like ZnO, which are not considered in the wurtzite ZnO structure. In addition, this study also presents a new perspective on the atomic behavior when a bulk crystal structure, which is not an original layered structure, is converted into an atomic-thick layered two-dimensional structure.

Keywords: graphene-like ZnO; atomic arrangement; merging; aberration-corrected TEM

1. Introduction

Bulk zinc oxide (ZnO) has a range of crystalline structures, namely wurtzite and zinc-blende structures [1]. The wurtzite ZnO is the most thermodynamically stable and exhibits strong electronic properties. It is also used in the flexible fabrication of electronic devices [2–5] and can transform into a variety of nanostructures [6]. When thinned down to a ZnO with atomic layers, planer ZnO [7–10], which resembles a graphene structure, can also form by expanding lattice parameter by 1.6% (a = 3.303 Å) [11] rather than the wurtzite ZnO. Since bulk ZnO is not a layered material, there is a limit to obtain atomic-thick nanosheets by physical and chemical exfoliation methods, such as methods related to graphene [12,13] and transition metal dichalcogenides [13–15]. However, it has been reported that ZnO nanosheets and graphene-like ZnO (g-ZnO) [7–10,16], which have trigonal planar coordination and are composed of Zn and O atoms alternately, can be synthesized by various deposition methods [7–9] with electron beam irradiation and a hydrothermal synthesis [10].

With the decreasing thickness of metal oxide semiconductor materials, unique electrical, mechanical, chemical, and optical properties are introduced, e.g., ZnO monolayer have increased band gap of ~4.0 eV comparing to bulk ZnO, which has a band gap of 3.37 eV, due to strong quantum confinement effects and graphene-like structure [8], while sustaining its direct band gap nature [17,18]. In addition, the g-ZnO is chemically stable [16,19,20] and is expected to exhibit high mechanical strength which originates from decreasing probability of finding defects by decreasing their thickness. In the same manner, low bending stiffness of the nanoscale thickness also enables g-ZnO to be applicable to the flexible electronics. Based on the extraordinary nature and properties of g-ZnO, it is very promising for transparent electronics, ultraviolet (UV) light emitters, chemical sensors, piezoelectric devices, switching electronics applications and photoactive devices [7,21,22], and ZnO QDs has potential applications in nanoscale devices [23].

We report the atomic arrangements at the edge of the g-ZnO nanosheets using aberration-corrected transmission electron microscopy (ACTEM). We explain the atomic reconstructions when the sheets merge at the edge of the g-ZnO nano-flakes. To observe that in detail, we use the g-ZnO nanosheets synthesized using an adaptive ionic layer epitaxy (AILE) method [24,25] and a monolayer graphene sheet as supporting layers inside transmission electron microscope (TEM). These results explain the atomic behavior of two-dimensional (2D) metal oxide semiconductors and direct observation of atomic arrangements from before to after merging.

2. Materials and Methods

2.1. Preparation of the Specimen

CVD-synthesized monolayer graphene film on 35 µm thick copper foil was purchased from Graphene Square (Graphene Square Inc., Seoul, Korea). Firstly, the graphene on copper foil was coated with ~200 nm poly methyl methacrylate (PMMA). Secondly, to release the graphene from the copper foil, the graphene on copper foil was floated onto copper etchant (Sigma Aldrich, St. Louis, MO, USA), then PMMA-coated graphene was transferred onto distilled water more than 3 times to remove the Cu etchant. The PMMA-coated graphene was then scooped up with TEM grids and annealed at 120 °C for 5 min. to firmly attach the PMMA-coated graphene onto the TEM grids. Lastly, the PMMA was dissolved with acetone for 1 day and the graphene on TEM grids were immersed into isopropyl alcohol several times to remove the acetone residue on the surface of the graphene. Synthesized ZnO nanosheets using the ILE [24,25] method were then scooped using the graphene-transferred TEM grid.

2.2. ARTEM Observations and STEM-EELS Spectra

Specimens were analyzed using an aberration-corrected FEI Titan Cubed TEM (FEI Titan3 G2 60–300), which was operated at an 80 kV acceleration voltage with a monochromator. The microscope provided a sub-Angstrom resolution at 80 kV and -13 ± 0.5 µm of spherical aberration (Cs). Typical electron beam densities were adjusted to ~6×10^5 e$^-$nm^{-2}. The atomic images were taken using a white atom contrast to obtain the actual atom positions under the properly focused conditions needed for direct image interpretation. STEM HAADF images and EELS spectra were recorded with a monochromatic beam at 80 kV with a probe size of 1.5 nm and an energy resolution of 0.8 eV, as measured from the full-width-at-half-maximum of the zero-loss peak.

3. Results and Discussion

We used synthesized g-ZnO nanosheets by an adaptive ionic layer epitaxy (AILE) method [24,25]. First, a monolayer graphene sheet is transferred on a TEM grid and then, the g-ZnO sheet is transferred on the monolayer graphene sheet (Figure 1a). We focused on the atomic arrangement and merging dynamics occurring from edge configurations of the ZnO nanosheets and nano-flakes. According to previous calculation and experimental research [7–9,16,17], wurtzite ZnO films can transform into a graphene-like structure, which is chemically stable. The wurtzite ZnO is schematically explained in Figure 1b. The Zn and O atoms form a tetrahedral configuration. G-ZnO has a hexagonal unit cell resembling a top-down view of the wurtzite ZnO structure in Figure 1b [26,27]. Since sp2 bonding of the hexagonal graphene-like structure is stronger than the sp3 bonding in the wurtzite structure, the Zn–O bond length of the g-ZnO structure is shorter as well [19–21].

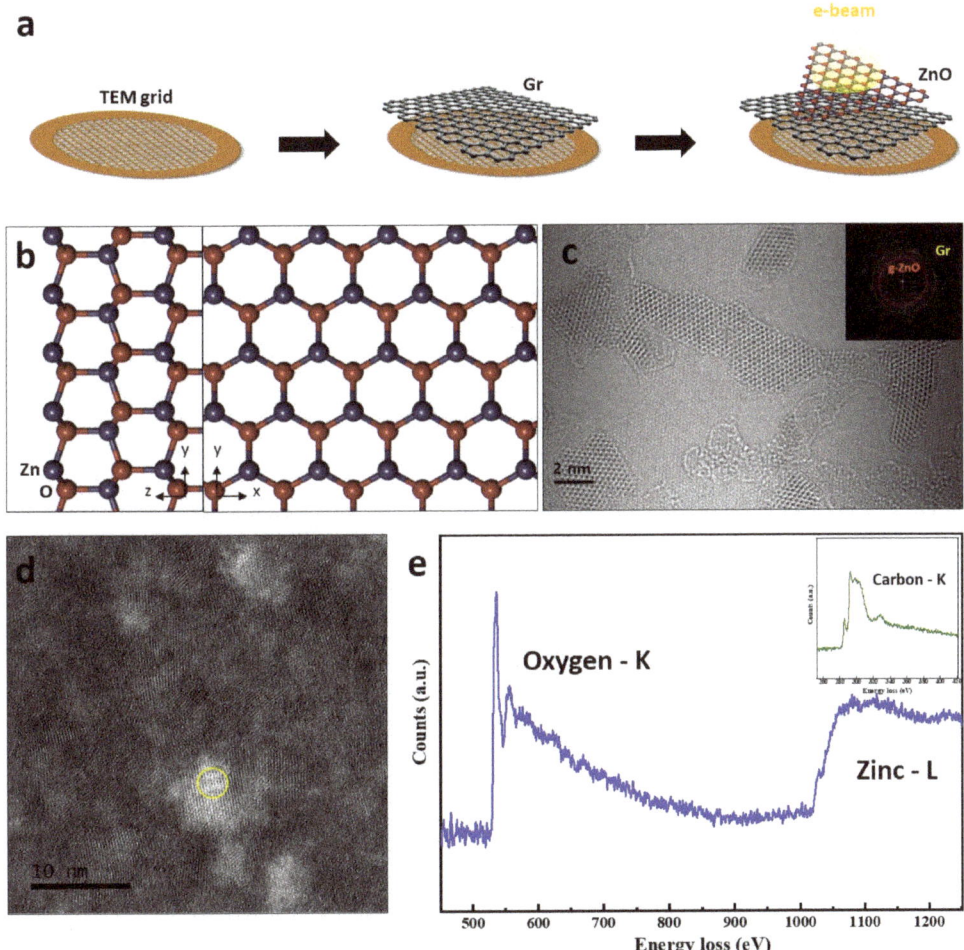

Figure 1. Graphene-like ZnO (g-ZnO) nanosheets on a monolayer graphene sheet. (**a**) Schematic showing a sample preparation for ACTEM imaging. (**b**) Atomic model showing top and side views of a wurtzite ZnO structure. (**c**) ACTEM image showing g-ZnO nanosheets on the monolayer graphene sheet with a FFT of the graphene and the ZnO sheets. (**d**) STEM HAADF image of the ZnO on the graphene. (**e**) EEL spectrum obtained from the yellow circle in (a) showing oxygen K-edge and zinc L-edge of the g-ZnO sheet on the graphene. The inset shows the carbon K-edge from the graphene.

For understanding the atomic movements of the g-ZnO nanosheets, we used the monolayer graphene sheets as a substrate layer because the ZnO sheets were not fully formed as lateral planar sheets. Since the lattice parameter of the ZnO and diameter of Zn atom is larger than their graphene, atomic dynamics of the ZnO sheets can be observed on the graphene sheet. An atomic image shows the g-ZnO sheets on the monolayer graphene sheets (as shown in Figure 1c) with an inset of fast Fourier transformation (FFT) showing spots of the graphene and spots of the ZnO sheets. Figure 1d,e shows a scanning transmission electron microscope high angle annular dark field (STEM-HAADF) image of a region where an electron energy loss (EEL) spectrum acquired and the EEL spectrum of the g-ZnO sheets on the graphene sheet, which confirms the presence of both Zn and O. The O-K edge where the peaks are located around 532 eV corresponds to oxygen atoms bonded to Zn in the ZnO nanosheet [28]. An inset of Figure 1d shows the C-K edge with π* and σ* peaks, which confirms the existence of graphene behind the ZnO sheets.

Figure 2 shows ACTEM images of edge configurations such as armchair (AC) and zigzag (ZZ) edge configurations nominally of the g-ZnO sheets. Previous research [8,9] shows that formation energies of O- and Zn-terminated ZZ edge configuration gradually decrease as the lateral size of the grown ZnO sheet increases at a lateral growth at the edge of the monolayer g-ZnO sheet. Furthermore, atomically extended ZZ edge configuration are observed (as shown in Figure 2a). For the AC edge configuration, although the lateral growth of ZnO is energetically favorable for the ZZ edge configurations rather that the AC edge configuration, the partial AC configuration is also able to be observed (as shown in Figure 2b).

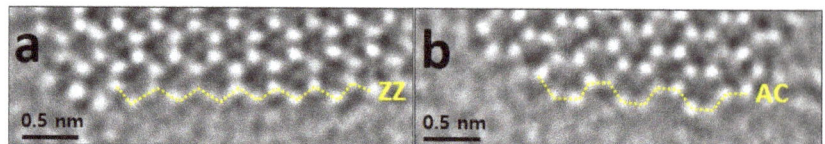

Figure 2. Edge configurations of g-ZnO. (**a**) ZZ edge configuration and (**b**) AC edge configuration.

In order to stabilize the edge of the synthetic g-ZnO sheet, but not the ZnO thinned from the bulky wurtzite ZnO, adatoms can be absorbed at the edge of the g-ZnO sheet. We observed that adatoms migrate along the edge of the ZnO sheet before being completely absorbed into the sheet (Figure 3). Although the migrating atoms at the edge might be carbons from carbon adsorbates and the graphene sheet, it is more likely that Zn atoms migrate because the Zn atom is clearly larger than the C. In addition, the positions in which the moving atoms settled are between O atoms designated at the O-terminated ZZ edge, which is confirmed using a line profile (as shown in Figure S1). Adatoms, colored with yellow dots, can attach and detach at the edge under continuous electron beam irradiation (Figure 3a,b). The attached adatom migrates freely along the O-terminated ZZ edge (Figure 3c–e) and finally separates from the ZZ edge (Figure 3f).

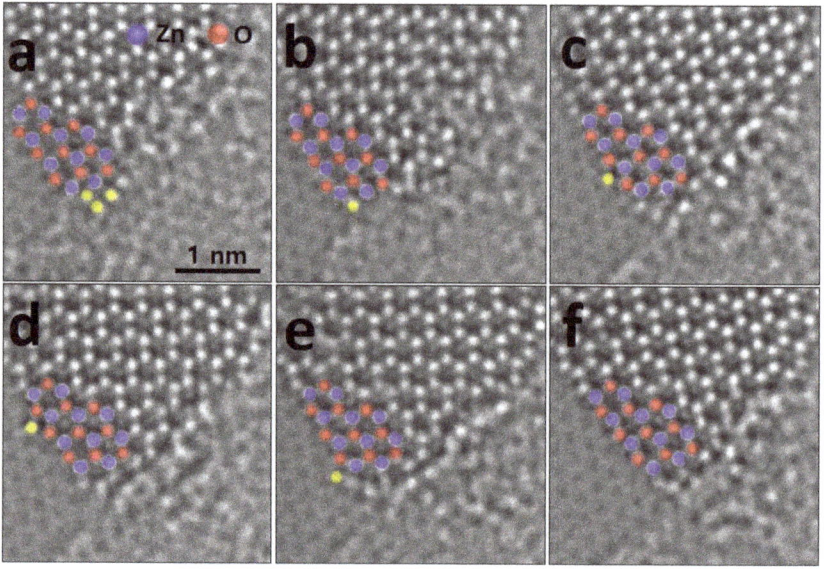

Figure 3. Atomic migration at the edge of the monolayer g-ZnO sheet. (**a,b**) successive images showing migrating adatoms, colored by yellow dots. (**c–e**) The adatom moves on top of dips in the ZZ configuration at the edge. (**f**) The adatom is finally detached.

About the lateral growth of the g-ZnO sheet, there are two directions which are parallel and normal directions to the growth direction (as shown in Figure 4a). Since the ZZ edge configuration is more stable than the AC, the normal direction to the growth direction is preferable to that parallel to the growth direction [8]. However, when a merging occurs from the edges of g-ZnO nano-flakes (Figure 4b), adatoms bond parallel to the flakes (Figure 4c). The flake correspondingly expands in the normal direction gradually (Figure 4d–f), maintaining a well-aligned atomical state (as shown in an inset of Figure 4f). The merged flake can also merge with others (Figure 4g), which continuously merges and expands through the parallel and normal directions from the flake (Figure 4h,i). Notably, when flakes merge, ZnO adatoms are, firstly, bonded along to the parallel direction. The adatoms are then bonded along to the normal direction. The hole inside the flake formed during the continuous merging and expanding process (Figure 4f) is filled by continuous adsorption of the adatoms through the parallel and normal direction indicating green and cyan arrows (Figure 4j,k). Finally, a fully merged flake is formed (Figure 4l).

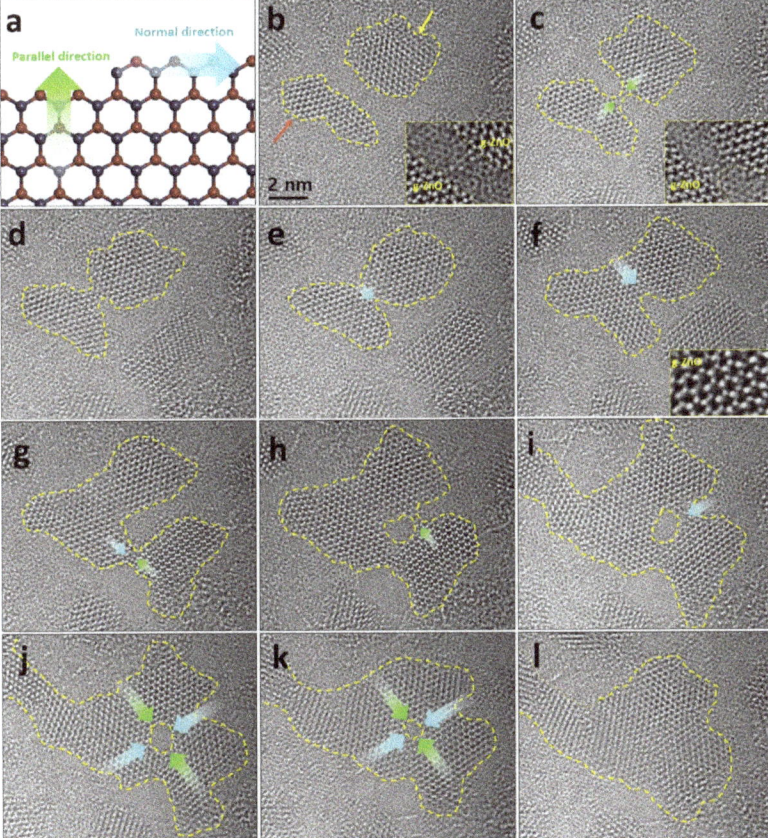

Figure 4. Merging of g-ZnO nano-flakes. (**a**) Atomic model showing normal and parallel to the growth direction. (**b**) Two graphene-like ZnO flakes before merging each other with magnified inset image showing edges of the flakes. Red and yellow arrows indicate that each flake is the same as flakes which are shown in Figure S2. (**c**) Initial state to merge the flakes along to the parallel directions of each flake. (**d–k**) Merging and expanding the flakes to the parallel and normal directions, indicated by green and cyan arrows. (**l**) Final g-ZnO flake after the process showing the behavior of merging and expanding.

We observed, under electron beam irradiation, atomic arrangement of bridged-adatoms when merging at edges of the g-ZnO. Figure 5 shows successive ACTEM images at the edges. This area was chosen for monitoring for its atomic arrangements facilitating direct atomic dynamics. In Figure 5a, each edge atomic configuration is confirmed using intensity line profiles at the flakes (see Figure S3). Before merging, Zn and O atoms that are not completely bound to the ZnO sheet on the graphene sheet may randomly exist around the edges of the nano-flakes. When the continuous electron beam is irradiated, the ZnO flakes merge, exhibiting partial rectangular lattices, colored by a yellow, as an intermediate state (Figure 5b–e). Finally, the bridged atoms rearrange into hexagonal lattices (Figure 5f).

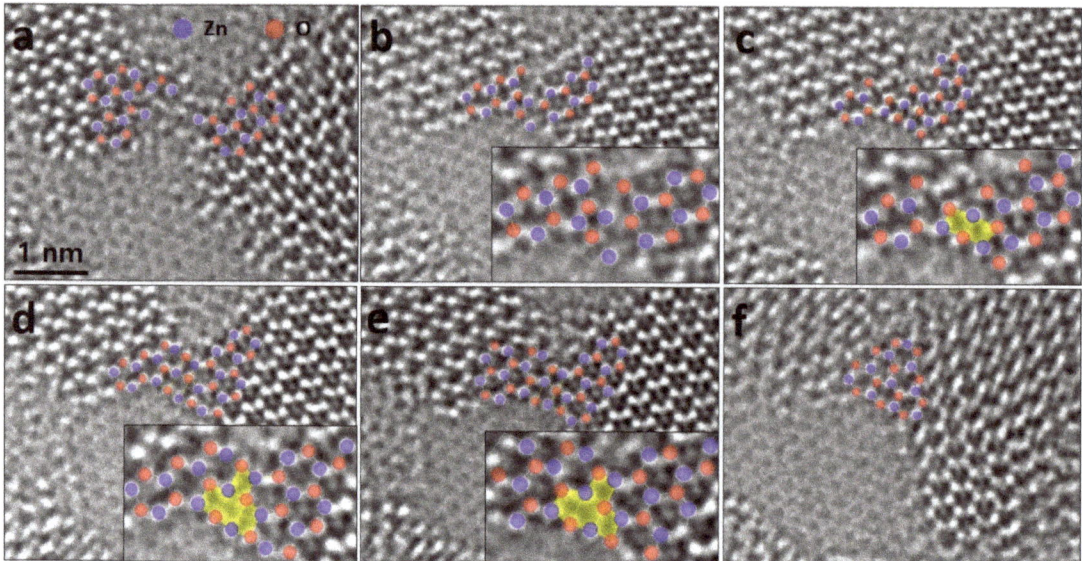

Figure 5. Atomic arrangement of bridged-atoms of the merging. (**a**) Initial state before the merging of the flakes. (**b–e**) Intermediate states showing the rectangular lattices with insets of the magnified bridged regions. (**f**) Final state after the merging. Raw images are supported as Figure S4.

4. Conclusions

We report the atomic behavior of the g-ZnO nanosheets on the graphene sheet under electron beam irradiations. Previous theoretical and experimental works suggest interesting properties for these sheets, but our work focuses on atomic arrangements of the edge and merging between the ZnO nano-flakes. The g-ZnO has edge configurations (e.g., the ZZ and AC configurations) which were analogous to the graphene lattices. In addition, we observed the atomic migration by adatoms at the edge of the sheet. In terms of merging of the g-ZnO sheets, when the nano-flakes merge, they have the preferred directions depending on the growth direction. Once adatoms bond along the parallel direction the edge, the flakes expand toward the normal direction. Moreover, while the merging occurs, intermediate arrangements (e.g., rectangular lattices) form before the perfect merging into the hexagonal lattices. This work explains the atomic dynamics of the promising g-ZnO nanosheet, which suggests fundamental investigations of the 2D metal oxide materials.

Supplementary Materials: The following are available online at https://www.mdpi.com/article/10.3390/nano11071833/s1, Figure S1: (a) Raw image of Figure 3e. (b) Intensity line profiles acquired from the red line in (a). This indicates that the ZnO flake has O-terminated ZZ configuration. Figure S2: Atomic g-ZnO flakes on the graphene. (a–i) ACTEM images showing the flakes atomically driven under electron beam irradiation. Figure S3: (a) Raw image of Figure 5a–c. Intensity line profiles acquired from blue and red arrows in (a). Figure S4: ACTEM images showing the full processes of the merging between the g-ZnO flakes.

Author Contributions: Conceptualization, G.H.R.; methodology, G.H.R. and J.C.Y.; formal analysis, G.H.R.; investigation, G.H.R. and J.C.Y. writing—original draft preparation, G.H.R.; writing—review and editing, G.H.R.; visualization, G.H.R. and J.C.Y.; supervision, G.H.R.; project administration, G.H.R.; funding acquisition, G.H.R. and Z.L. All authors have read and agreed to the published version of the manuscript.

Funding: This research was supported by Korea government (MSIT) (No. 2020R1G1A1099542) and Institute for Basic Science (IBS-R019-D1).

Data Availability Statement: The data presented in this study are available in [insert article or Supplementary Materials here].

Acknowledgments: This work was supported by the National Research Foundation of Korea (NRF) grant funded by the Korea government (MSIT) (No. 2020R1G1A1099542) and Institute for Basic Science (IBS-R019-D1).

Conflicts of Interest: The authors declare no conflict of interest.

References

1. Morkoc, H.; Ozgur, U. *Zinc Oxide: Fundamentals, Materials and Device Technology*; Wiley-VCH: Weinheim, Germany, 2009.
2. Choi, D.; Choi, M.Y.; Choi, W.M.; Shin, H.J.; Park, H.K.; Seo, J.S.; Park, J.; Yoon, S.M.; Chae, S.J.; Lee, Y.H.; et al. Fully Rollable Transparent Nanogenerators Based on Graphene Electrodes. *Adv. Mater.* **2010**, *22*, 2187–2192. [CrossRef]
3. Chung, K.; Lee, C.H.; Yi, G.C. Transferable GaN Layers Grown on ZnO-Coated Graphene Layers for Optoelectronic Devices. *Science* **2010**, *330*, 655–657. [CrossRef]
4. Lee, C.H.; Kim, Y.J.; Hong, Y.J.; Jeon, S.R.; Bae, S.; Hong, B.H.; Yi, G.C. Flexible Inorganic Nanostructure Light-Emitting Diodes Fabricated on Graphene Films. *Adv. Mater.* **2011**, *23*, 4614–4619. [CrossRef]
5. Fan, Z.; Lu, J.G. Zinc Oxide Nanostructures: Synthesis and Properties. *J. Nanosci. Nanotechnol.* **2005**, *5*, 1561–1573. [CrossRef]
6. Wang, Z.L. Zinc Oxide Nanostructures: Growth, Properties and Applications. *J. Phys. Condens. Matter* **2004**, *16*, 829–858. [CrossRef]
7. Ta, H.Q.; Bachmatiuk, A.; Dianat, A.; Ortmann, F.; Zhao, J.; Warner, J.H.; Eckert, J.; Cunniberti, G.; Rümmeli, M.H. In-Situ Observations of Freestanding Graphene-Like Mono- and Bi-layer ZnO Membranes. *ACS Nano* **2015**, *9*, 11408–11413.
8. Hong, H.-K.; Jo, J.; Hwang, D.; Lee, J.; Kim, N.Y.; Son, S.; Kim, J.H.; Jin, M.-J.; Jun, Y.C.; Erni, R.; et al. Atomic Scale Study on Growth and Heteroepitaxy of ZnO Monolayer on Graphene. *Nano Lett.* **2017**, *17*, 120–127. [CrossRef] [PubMed]
9. Son, S.; Cho, Y.; Hong, H.-K.; Lee, J.; Kim, J.H.; Kim, K.; Lee, Y.; Yoon, A.; Shin, H.-J.; Lee, Z. Spontaneous Formation of a ZnO Monolayer by the Redox Reaction of Zn on Graphene Oxide. *ACS Appl. Mater. Interfaces* **2020**, *12*, 54222–54229. [CrossRef] [PubMed]
10. Sahoo, T.; Nayak, S.K.; Chelliah, P.; Rath, M.K.; Prida, B. Observations of Two-Dimensional Monolayer Zinc Oxide. *Mater. Res. Bull.* **2016**, *75*, 134–138. [CrossRef]
11. Tusche, C.; Meyerheim, H.L.; Kirschner, J. Observation of Depolarized ZnO(0001) Mono-Layers: Formation of Unreconstructed Planar Sheets. *Phys. Rev. Lett.* **2007**, *99*, 026102. [CrossRef] [PubMed]
12. Yi, M.; Shen, Z. A Review on Mechanical Exfoliation for the Scalable Production of Graphene. *J. Mater. Chem. A* **2015**, *3*, 11700–11715. [CrossRef]
13. Huo, C.; Yan, Z.; Song, X.; Zeng, H. 2D Materials via Liquid Exfoliation: A Review on Fabrication and Applications. *Sci. Bull.* **2015**, *60*, 1994–2008. [CrossRef]
14. Eng, A.Y.S.; Ambrosi, A.; Sofer, Z.; Šimek, P.; Pumera, M. Electrochemistry of Transition Metal Dichalcogenides: Strong Dependence on the Metal-to-Chalcogen Composition and Exfoliation Method. *ACS Nano* **2014**, *8*, 12185–12198. [CrossRef]
15. Chhowalla, M.; Liu, Z.; Zhang, H. Two-Dimensional Transition Metal Dichalcogenide (TMD) Nanosheets. *Chem. Soc. Rev.* **2015**, *44*, 2584–2586. [CrossRef] [PubMed]
16. Ta, H.Q.; Zhao, L.; Pohi, D.; Pang, J.; Trzebicka, B.; Rellinghaus, B.; Pribat, D.; Gemming, T.; Liu, Z.; Bachmatiuk, A.; et al. Graphene-Like ZnO: A Mini Review. *Crystals* **2016**, *6*, 100. [CrossRef]
17. Topsakal, M.; Cahangirov, S.; Bekaroglu, E.; Ciraci, S. First-Principles Study of Zinc Oxide Honeycomb Structures. *Phys. Rev. B* **2009**, *80*, 235119. [CrossRef]
18. Tu, Z.C. First-Principles Study on Physical Properties of a Single ZnO Mono-Layer with Graphene-Like structure. *J. Comput. Theor. Nanosci.* **2010**, *7*, 1182–1186. [CrossRef]
19. Claeyssens, F.; Freeman, C.L.; Allan, N.L.; Sun, Y.; Ashfolda, M.N.R.; Harding, J.H. Growth of ZnO Thin Films-Experiment and Theory. *J. Mater. Chem.* **2005**, *15*, 139–148. [CrossRef]
20. Tu, Z.C.; Hu, X. Elasticity and Piezoelectricity of Zinc Oxide Crystals, Single Layers, and Possible Single-Walled Nanotubes. *Phys. Rev. B* **2006**, *74*, 035434. [CrossRef]

21. Lee, J.; Sorescu, D.C.; Deng, X.Y. Tunable Lattice Constant and Band Gap of Single- and Few-Layer ZnO. *J. Phys. Rev. Lett.* **2016**, *7*, 1335–1340. [CrossRef]
22. Demel, J.; Plestil, J.; Bezdicka, P.; Janda, P.; Klementova, M.; Lang, K. Few-Layer ZnO Nanosheets: Preparation, Properties, and Films with Exposed {001} Facets. *J. Phys. Chem. C* **2011**, *115*, 24702–24706. [CrossRef]
23. Liu, G.; Debnath, B.; Pope, T.R.; Salguero, T.T.; Lake, R.K.; Balandin, A.A. A Charge-Density-Wave Oscillator Based on an Integrated Tantalum Disulfide-Boron Nitride-Graphene Device Operating at Room Temperature. *Nat. Nanotechnol.* **2016**, *11*, 845–850. [CrossRef]
24. Wang, F.; Yin, X.; Wang, X. Morphological Control in the Adaptive Ionic Layer Epitaxy of ZnO Nanosheets. *Extrem. Mech. Lett.* **2016**, *7*, 64–70. [CrossRef]
25. Wang, F.; Seo, J.-H.; Luo, G.; Starr, M.B.; Li, Z.; Geng, D.; Yin, X.; Wang, S.; Fraser, D.G.; Morgan, D.; et al. Nanometre-Thick Single-Crystalline Nanosheets Grown at the Water–Air Interface. *Nat. Commun.* **2016**, *7*, 10444. [CrossRef]
26. Chai, G.L.; Lin, C.S.; Cheng, W.D. First-Principles Study of ZnO Cluster-Decorated Carbon Nanotubes. *Nanotechnology* **2011**, *22*, 445705. [CrossRef] [PubMed]
27. Pandey, D.K.; Yadav, P.S.; Agrawal, S.; Agrawal, B.K. Structural and Electronic Properties of ZnO Nanoclusters: A B3LYP DFT Study. *Adv. Mater. Res.* **2013**, *650*, 29–33. [CrossRef]
28. Sakaguchi, N.; Suzuki, Y.; Watanabe, K.; Iwama, S.; Watanabe, S.; Ichinose, H. A HRTEM and EELS Study of Pd/ZnO Polar Interfaces. *Philos. Mag.* **2008**, *88*, 1493–1509. [CrossRef]

Review

2D Material and Perovskite Heterostructure for Optoelectronic Applications

Sijia Miao [1], Tianle Liu [1], Yujian Du [2], Xinyi Zhou [1], Jingnan Gao [1], Yichu Xie [1], Fengyi Shen [1], Yihua Liu [1] and Yuljae Cho [1,*]

[1] UM-SJTU Joint Institute, Shanghai Jiao Tong University, Shanghai 200240, China; sijia_miao@sjtu.edu.cn (S.M.); tyrion_l2017@sjtu.edu.cn (T.L.); zxylily@umich.edu (X.Z.); gjn0310@sjtu.edu.cn (J.G.); xieyichu@umich.edu (Y.X.); 1517511250@sjtu.edu.cn (F.S.); ayka_tsuzuki@sjtu.edu.cn (Y.L.)

[2] School of Microelectronics, Dalian University of Technology, Dalian 116620, China; imduke@mail.dlut.edu.cn

* Correspondence: yuljae.cho@sjtu.edu.cn

Abstract: Optoelectronic devices are key building blocks for sustainable energy, imaging applications, and optical communications in modern society. Two-dimensional materials and perovskites have been considered promising candidates in this research area due to their fascinating material properties. Despite the significant progress achieved in the past decades, challenges still remain to further improve the performance of devices based on 2D materials or perovskites and to solve stability issues for their reliability. Recently, a novel concept of 2D material/perovskite heterostructure has demonstrated remarkable achievements by taking advantage of both materials. The diverse fabrication techniques and large families of 2D materials and perovskites open up great opportunities for structure modification, interface engineering, and composition tuning in state-of-the-art optoelectronics. In this review, we present comprehensive information on the synthesis methods, material properties of 2D materials and perovskites, and the research progress of optoelectronic devices, particularly solar cells and photodetectors which are based on 2D materials, perovskites, and 2D material/perovskite heterostructures with future perspectives.

Keywords: 2D materials; perovskites; heterostructures; optoelectronics

1. Introduction

As modern technology improves by leaps and bounds, optoelectronic devices, such as solar cells and photodetectors, have become indispensable parts of society. These semiconductor devices that convert optical signals into electrical ones have wide applications in the areas of renewable energy, imaging systems, biomedical devices, and optical communications [1–4]. Over the past few decades, tremendous efforts have been made persistently to enhance the efficiency and stability of the devices through synthesizing high-quality materials [5–12], designing novel device structures, engineering interfaces [13], and employing encapsulation with polymer and inorganic glass [14,15].

2D materials, including graphene and its derivatives, transition metal dichalcogenides (TMDCs), and black phosphorous (BP), have attracted intense interest in optoelectronic applications due to their distinctive optical, electrical, and mechanical characteristics [12,16–20]. These nanomaterials possess strong in-plane covalent bonds whereas each layer is vertically connected by weak van der Waals (vdWs) force. The unique vdWs structures enable the exfoliation of an ultrathin single layer with a uniformly distributed thickness and orientation [21–24]. Despite the great potential of 2D materials in advanced optoelectronics, their atomic-scale thickness largely restricts the light absorption capacity of devices, resulting in unsatisfactory device performance [25].

Meanwhile, perovskites have been regarded as a promising active layer in high-performance optoelectronic devices. Since the first demonstration of perovskite solar cells

with a power conversion efficiency (PCE) of 3.8% in 2009 [26], the record PCE has been elevated to over 25% till now [27], almost catching up with commercial photovoltaic applications [26,27]. Similarly, perovskite-based photodetectors have also achieved impressive progress, with the champion device exhibiting detectivity exceeding 10^{14} Jones and fast response speed at the nanosecond level [28,29]. The outstanding performance of perovskite devices mainly stemmed from their fascinating properties, such as high light absorption coefficient, adjustable bandgap, and long charge carrier diffusion length [30,31]. However, challenges still remain regarding the long-term stability issues and unsatisfactory interfaces, which require further intensive research [5,6,32].

Recently, the integration of 2D materials and perovskites for heterostructures has stimulated a research hotspot. The heterostructures enable the absorption of more photons and enhanced resistance to moisture and oxygen by combining the advantages of both materials, bringing more opportunities for future optoelectronic industries. In addition, some 2D materials are found to be superior alternatives for carrier transport layers [13,33]. In light of this, it is of importance to review comprehensive aspects including the materials, devices, and integration technology of 2D materials and perovskites.

In this regard, we present comprehensive information on the applications of 2D materials, perovskites, and 2D material/perovskite heterostructures for applications to solar cells and photodetectors. We begin with an introduction to the synthesis methods and basic properties of 2D materials and perovskites. Then, we summarize the development of optoelectronic devices based on 2D materials and perovskites and address the existing issues. Further, we introduce the newly-emerged concept of 2D material/perovskite heterostructure and its impressive progress in optoelectronics. We focus on the synergistic effects of integrating these materials and the underlying mechanisms of performance enhancement. Finally, we discuss the remaining challenges of developing stable, high-quality, and environmentally friendly 2D material/perovskite heterostructures.

2. Synthesis of 2D Materials

2.1. Graphene and Its Derivatives

Graphene is one atomic layer of graphite with a thickness of approximately 0.35 nm and has attracted lots of interest from academia and industries due to its excellent properties, such as a theoretical specific surface area (2630 m^2 g^{-1}), optical transparency, intrinsic mobility (200,000 cm^2 V^{-1} s^{-1}), Young's modulus (~1.0 TPa) and thermal conductivity (~5000 W m^{-1} K^{-1}) [12,16,17,34]. Ever since its first demonstration, tremendous efforts have been put into its synthesis [21]. Among various methods to synthesize graphene, we will introduce mechanical exfoliation, liquid exfoliation, and chemical vapor deposition (CVD), which are commonly used.

2.1.1. Mechanical Exfoliation

Graphene was believed not to exist in the free state until Geim and Novoselov obtained it via mechanical exfoliation in 2004 [21]. The thin flakes of graphite could be successively cleaved into thinner flakes using adhesive tapes down to a single atomic layer (Figure 1a). The crystallinity of the obtained flakes was high and further examinations confirmed the anomalous quantum Hall effect which was predicted by Semenoff [35–37]. These observations not only showed the great potential of graphene as a platform for studying quantum physics, kicking off the race in this field but also yielded a Nobel prize. However, the mechanical exfoliation was inefficient in producing a large volume of graphene and the size obtained was limited.

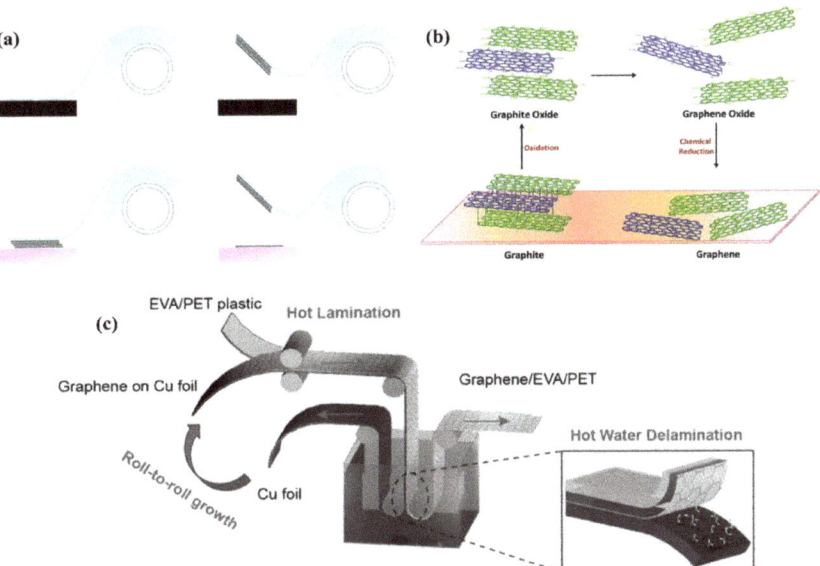

Figure 1. (a) Schematics of mechanical exfoliation method [38]; (b) Schematic of chemical exfoliation method [39]; (c) Schematic of roll-to-roll production of graphene from a Cu foil to a target substrate [40]. (a) Reprinted with permission from Ref. [38]. Copyright 2011, WILEY. (b) Reprinted with permission from Ref. [39]. Copyright 2014, RSC. (c) Reprinted with permission from Ref. [40]. Copyright 2015, WILEY.

2.1.2. Chemical Exfoliation

Since graphene tends to aggregate in commonly available solvents, researchers have developed the chemical exfoliation method. In this method, graphite is first oxidized to introduce hydrophilic functional groups, such as carboxyl, hydroxyl, and epoxide groups, and then dispersed in water via sonication. The dispersed solution is finally reduced to graphene by thermal annealing or by using chemical reagents (Figure 1b) [39]. Therefore, graphene synthesized in this way is termed reduced graphene oxide (rGO). Typically, $KMnO_4$, $NaNO_3$, and a strong acid (e.g., H_2SO_4) are used as an oxidizer. This synthesis route developed by Hummers is safer compared with $KClO_3$ used initially. Later, other researchers modified Hummers' method by using various reduction agents, such as hydrohalic acids, aluminum hydride, and sodium ammonia [41–43]. Chemical exfoliation is cost-effective and scalable. However, the method introduces defects in graphene in the oxidation process, which degrades properties such as electron mobility and thermal conductivity.

2.1.3. Chemical Vapor Deposition

The CVD of graphene is typically performed on copper or nickel through a thermal method. Methane is a commonly used carbon source. As it is exposed to the heated metal surface, the metal catalyzes the loss of hydrogen and dissolves the remaining carbon, leading to a layer of metal carbide [12]. As the temperature drops, the metal carbide surface layers saturate, and graphitic carbon precipitates out. In this way, large-area single-layer graphene can be obtained. Though researchers have known for a long time that the CVD of hydrocarbon on metals can generate thin graphite layers, the breakthrough of CVD for graphene did not occur until self-limiting graphene growth on Cu was discovered in 2009 [44]. Since the graphene growth was irrelevant to the thickness of Cu foil and C solubility in Cu was very low, the proposed mechanism for the self-limiting growth was surface-catalysis rather than precipitation during cooling. In addition, patterned graphene could be obtained by patterning the metal substrate used for growth, which offers huge benefits in patterning graphene after the synthesis [45]. Remarkably, a large area of graphene was also successfully attained by the roll-to-roll process for flexible transparent electrodes (Figure 1c) [40,46].

2.2. 2D Transition Metal Dichalcogenides

2D transition metal dichalcogenides (TMDCs), such as MoS_2, $MoSe_2$, and WS_2, have gained great attention because they show properties different from their bulk counterparts [10]. For example, the MoS_2 monolayer is a direct bandgap semiconductor while the bulk crystal is an indirect one with a narrower bandgap. Thus, monolayer MoS_2 shows much stronger photoluminescence than the multiple layers. Among various methods, we will introduce mechanical exfoliation, liquid exfoliation, and chemical vapor deposition, which are the most widely used methods to synthesize 2D TMDCs [22,23,47–49].

2.2.1. Mechanical Exfoliation

After the successful demonstration of mechanical exfoliation of graphene, it has been extended to 2D TMDCs as well. Typically, an adhesive tape is used to peel off a thin layer from a bulk crystal, and then the layer is transferred to a target substrate, leaving single or multiple layers of TMDCs on the substrate [38,50]. Although mechanical exfoliation provides nanosheets with clean surfaces and a good crystallinity, it is not scalable and does not have systematic control over layer size and thickness. The layers mechanically exfoliated are typically of several microns size. To obtain larger layers, Magda and coworkers employed the chemical affinity of S atoms to Au to achieve MoS_2 single layers of several hundred microns size on Au (111) surfaces [22]. Furthermore, they found that their method could be extended to other chalcogenides, such as WSe_2 and Bi_2Te_3 as well.

2.2.2. Liquid Exfoliation

Liquid exfoliation is another popular method to fabricate 2D TMDCs which can be further categorized as solvent intercalation, chemical intercalation, and electrochemical intercalation (Figure 2a,b) [49,51,52]. It typically yields nanosheets with small lateral sizes (<1 µm) which tend to aggregate upon deposition. The main idea behind the method is to weaken the vdWs interaction between layers to separate them. Organolithium is commonly used for chemical intercalation (Figure 2a). However, in the case of MoS_2, the as-exfoliated nanosheets showed a dominant metastable metallic phase rather than a semiconducting phase due to Li intercalation. Eda and coworkers found that mild annealing at 300 °C led to the gradual restoration of the semiconducting phase [20].

It is also viable to exfoliate single TMDC layers through direct sonication in organic solvents or solvent intercalation (Figure 2a). Coleman et al. tested various organic solvents in many TMDCs (MoS_2, $MoSe_2$, WS_2, etc.) and found N-methyl-pyrrolidone (NMP) and isopropanol (IPA) resulted in stable dispersions [53]. However, the yield was low and the dispersion also contained multilayer TMDCs.

In order to improve the yield and obtain better control over the intercalation process, Zeng et al. developed the electrochemical intercalation method (Figure 2b) [49]. A bulk TMDC was incorporated as a cathode of a test cell while a Li foil was used as an anode to provide intercalation ions. Hence, the degree of intercalation could be controlled in the discharge process, leading to a 92% yield of single-layer TMDC.

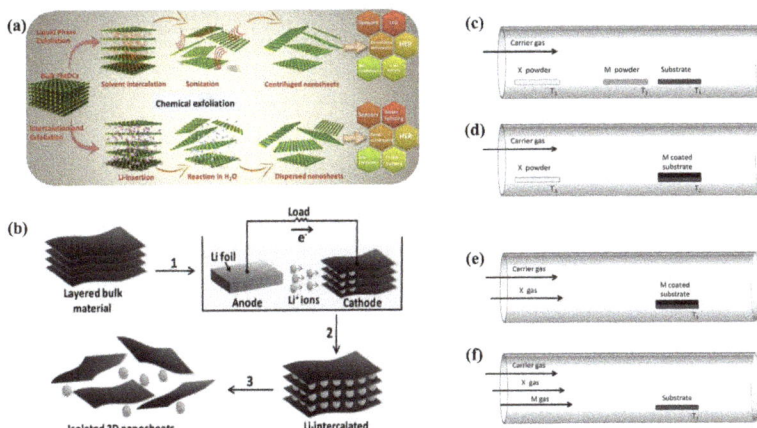

Figure 2. (a,b) Different liquid exfoliation mechanisms: (a) solvent intercalation, and chemical intercalation [52], and (b) electrochemical intercalation [49]; (c–f) Chemical vapor deposition setups depending on the precursors [47]. (a) Reprinted with permission from Ref. [52]. Copyright 2021, WILEY. (b) Reprinted with permission from Ref. [49]. Copyright 2019, RSC. (c–f) Reprinted with permission from Ref. [47]. Copyright 2015, RSC.

2.2.3. Chemical Vapor Deposition

CVD is the most widely used bottom-up technology in fabricating atomically thin TMDCs. As a well-developed technique, it has better control of synthesis parameters and is, therefore, able to produce large-area monolayer TMDCs. Various precursors can be used as metal and chalcogen sources, such as MoO_3, WO_3, Se powder, and H_2S [47,54–56]. For the CVD method, a different experimental setup needs to be adopted depending on the types of sources (Figure 2c–f). Figure 2c shows the setup for metal and chalcogen sources in powders. Since evaporation of metal (or metal oxides) and synthesis of TMDC require a temperature much higher than the boiling point of chalcogen powders, it is important to control the furnace temperature profile. When the metal has a very high boiling temperature (e.g., tungsten), its oxide (e.g., WO_3) is usually used instead [55]. When Se is used, H_2 in the carrier gas is usually required as a reducing agent [56]. During the synthesis, seeding promoters can also be applied to avoid unwanted structures, such as nanorods and nanoparticles.

TMDC monolayers can also be synthesized through the chalcogenization of metal or metal oxide deposited on the substrate (Figure 2d,e). The thickness of the metal or metal oxide layer is about 1–5 nm and determines the number of TMDC layers [57]. Hence, the difficulty of this method lies in the precise control of thickness and homogeneity of the pre-deposited layer over a large area. Figure 2f illustrates the setup when both metal and chalcogenide sources are in the gas phase. Common precursors are $Mo(CO)_6$, $W(CO)_6$, H_2S, and dimethyl disulfide (DMDS) [47]. In the case of $Mo(CO)_6$ and dimethyl disulfide (DMDS) forming MoS_2, the reaction can happen at low temperatures (100–140 °C), but the resulted nanosheets require subsequent annealing at 900 °C to obtain a better crystallinity [58].

3. Synthesis of Lead Halide Perovskites

Lead halide perovskites (LHPs) with a chemical formula ABX_3 (B = Pb and X = Cl, Br, or I) are promising materials in optoelectronic applications due to their high defect tolerance, large light absorption coefficients, and long charge carrier diffusion length. According to the requirement for ionic radii, there are only three suitable cations in cubic α-phase, namely, cesium (Cs^+), methylammounium (MA^+), and formamidium (FA^+) ions. These materials have been applied in various optoelectronics such as solar cells, light-emitting diodes (LEDs), and photodetectors [59–61]. However, bulk perovskites are unstable under ambient atmosphere and irradiation, which triggers studies on their more stable reduced-dimensional counterparts, namely 0D quantum dots, 2D, and quasi-2D. Comprehensive reviews on the synthesis of perovskites can be found in [5–9,62,63], and therefore we summarize the synthesis methods of perovskites with different dimensions for optoelectronic applications in this review.

3.1. 3D Lead Halide Perovskites

For 3D LHPs, tremendous efforts have been made in engineering materials' stability and morphology in order to fully utilize outstanding materials properties which were overshadowed by the poor air stability and film morphology. Among various film deposition methods, here we focus on the most common approaches, namely, two-step sequential deposition, one-step spin-coating, and thermal evaporation (Figure 3a,b). Various strategies have been explored in the frame of those methods to stabilize LHPs and improve the quality of films, such as mixing A-site cations and X-site anions, and adding organic molecules or lower dimensional LHP to passivate surfaces and grain boundaries [61,64,65].

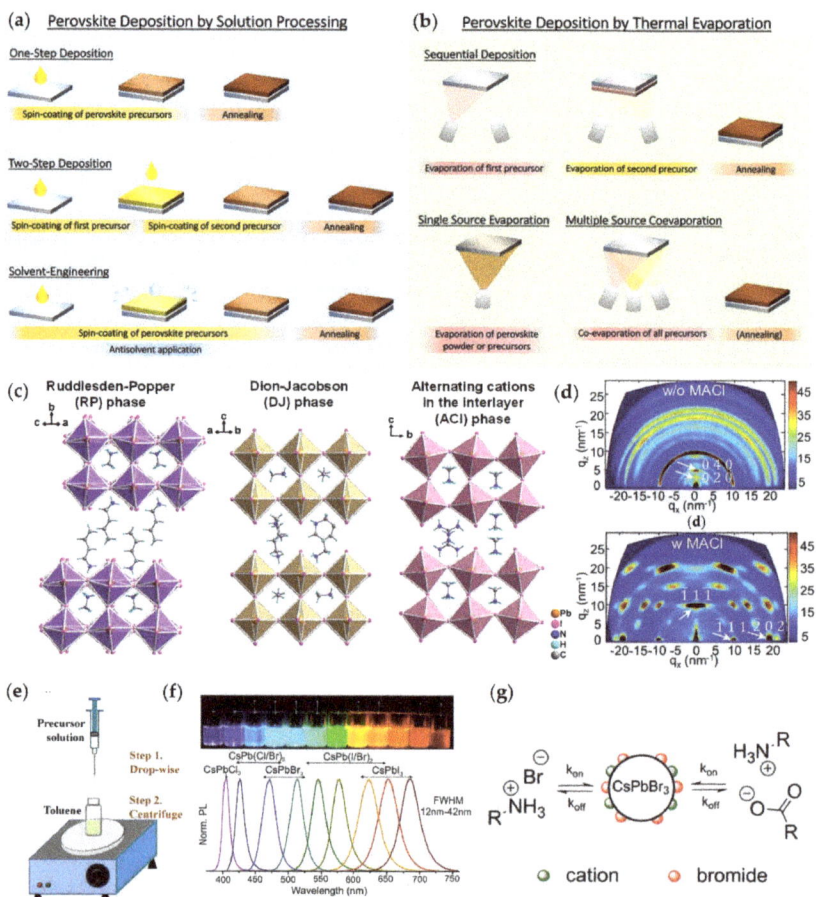

Figure 3. (**a**) Schematic illustration of one-step, two-step sequential, and antisolvent method for 3D LHP film deposition [63]; (**b**) Schematic illustration of single-source, sequential, and multi-source thermal evaporation [63]; (**c**) Examples showing structures of RP, DJ, and ACI phases of <100>-oriented LHP respectively [7]; (**d**) Grazing-incident wide-angle X-ray scattering patterns of

(PXD)(MA)$_2$Pb$_3$I$_{10}$ with and without MACl additive [66]; (**e**) Schematic presentation of the LARP method [67]; (**f**) CsPbX$_3$ PQDs in toluene under UV lamp (365 nm) and the normalized PL spectra [68]; (**g**) Schematic representation of the dynamic ligand binding of OA and OAm to PQD surfaces [69]. (**a**,**b**) Reprinted with permission from Ref. [63]. Copyright 2020, WILEY. (**c**) Reprinted with permission from Ref. [7]. Copyright 2021, ACS. (**d**) Reprinted with permission from Ref. [66]. Copyright 2020, WILEY. (**e**) Reprinted with permission from Ref. [70]. Copyright 2015, ACS. (**f**) Reprinted with permission from Ref. [68]. Copyright 2015, ACS. (**g**) Reprinted with permission from Ref. [69]. Copyright 2016, ACS.

The two-step sequential deposition was one of the widely used methods in the early age of perovskite research for 3D LHP films [63]. Typically, an inorganic layer, for example, PbI$_2$, is first spin-coated on a substrate, and then the deposited inorganic layer is immersed into an organic salt solution, for example, methylammonium iodide (MAI). Alternatively, the organic salt solution is spin-coated on the inorganic layer to form the desired 3D LHPs (Figure 3a). For the two-step sequential deposition methods, it is important to control parameters to obtain the high-quality film, such as the morphology of the inorganic layer at the first step, the concentration of the organic salt solution, and the speed of the spin-coating [62]. Although the two-step method offers good reproducibility and compact films, it often encounters an issue of incomplete conversion. This has resulted from the formation of perovskite on the surface of the inorganic layer which inhibits the diffusion of organic cations. Gong et al. introduced nickel chloride (NiCl$_2$) as an additive in the PbI$_2$ precursor to form a porous film [71]. As a result, the MAI solution could penetrate the PbI$_2$ film more easily, leading to the formation of larger perovskite grains. Engineering solvents have also been introduced as another strategy. Zhi and coworkers added *N*-Methyl-2-pyrrolidone (NMP) with low volatility to the MAI precursor to assist the recrystallization process of the 3D LHPs and to enhance Ostwald ripening, resulting in films with large columnar grains [72].

3D LHP films can also be fabricated via a one-step spin-coating method. In this method, all the components are dissolved in the same solvent to form a precursor and spin-coated generally at two stages, such as at a low speed for a short time and then at a high speed for a long time (Figure 3a). However, the resultant films could have pinholes, rough surfaces, and small grains. To resolve these issues, additives can incorporated in precursors before spin-coating. Lewis acids (e.g., iodopentafluorobenzene) or bases (e.g., methylammonium chloride (MACl)) are commonly used to passivate under-coordinated halide or Pb atoms at grain boundaries or surfaces, while other additives like Pb(SCN)$_2$ are shown to improve the crystallinity and grain size of 3D LHPs [5,73,74]. Another approach is to drip antisolvent on substrates at the second stage of spin-coating to facilitate homogeneous nucleation by extracting the host solvent (Figure 3a). A wide range of antisolvent has been employed, from highly polar ethyl acetate to nonpolar toluene, from diethyl ether with a low boiling point to chlorobenzene with a high boiling point [75–78]. According to Taylor and coworkers, the application of the antisolvent is not instantaneous and its duration is also an important factor [79].

Other than the spin-coating method, thermal evaporation is another film deposition method widely employed. The evaporation method can be further categorized as a single-source, sequential, and multisource co-deposition (Figure 3b) [63]. In the single-source thermal evaporation, either the prepared perovskite powder or the mixed precursor is put in one crucible for the thermal evaporation. In the sequential evaporation, the individual precursor is evaporated separately, followed by a thermal or vapor annealing. The most common method is the multisource co-evaporation where each precursor is placed in different crucibles and evaporated at the same time. The thermal evaporation method is suitable for large-scale fabrication and conformal films. It also allows precursors with solvent orthogonality and prevents damaging underlying layers during deposition. However, there are drawbacks to thermal evaporation as well. It requires expensive and complex vacuum systems, which limits the number of studies. The grain size of thermally evaporated films is typically smaller than that of the solution-processed films. Lohmann et al. found that as the substrate temperature decreased from room temperature to −2 °C, the grain size of MAPbI$_3$ increased from 100 nm to micrometer size [80]. However, it has also been pointed out that the evaporation of MAI is difficult to control due to its relatively high vapor pressure and low decomposition temperature [81]. To solve this problem, vapor-deposited MA-free perovskite films have been studied. For instance, Lohmann and coworkers showed that partially substituting PbCl$_2$ for PbI$_2$ could significantly suppress defects in FA$_{0.83}$Cs$_{0.17}$PbI$_3$ films [82].

The deposition methods mentioned above aim at fabricating films with uniform and full coverage. However, sometimes localized crystallization and patterned films are preferred. A laser-assisted crystallization method shows advantages in this sense. Arciniegas et al. demonstrated confined growth of MAPbBr$_3$ via laser irradiation [83]. The growth mechanism was ascribed to the local generation of MA ions from the N-methylformamide solvent due to laser-induced heat. In addition to crystal growth, laser irradiation could also lead to a reversible localized phase transition. Zou and coworkers illustrated a nonvolatile rewritable photomemory array based on this principle [84].

3.2. 2D and Quasi-2D Lead Halide Perovskites

(Quasi-)2D LHPs can be seen as a slice cleaved from an ideal cubic perovskite along certain crystalline planes, such as <100> by layers of organic ions [7,85]. Therefore, they can be categorized according to the cleaving planes and the thickness of the inorganic layers between the organic spacers. <100>-oriented (quasi-)2D LHPs have a chemical formula A$'_2$A$_{n-1}$Pb$_n$X$_{3n+1}$ or A$'$A$_{n-1}$Pb$_n$X$_{3n+1}$ (A$'$ = 1+ or 2+, A = 1+ cation, X = Cl, Br, or I) and can be further classified as: (i) Ruddlesden-Popper (RP) phase, (ii) Dion-Jacobson (DJ) phase, and (iii) alternating cation in the interlayer space (ACI) phase (Figure. 3c). All three phases consist of inorganic layers of corner-sharing [PbX$_6$]$^{4-}$ octahedrons. The difference between these phases lies in the stacking displacement of adjacent layers. The RP phase is characterized by the interdigitated organic bilayer which causes the inorganic layers staggered by half a unit cell (0.5, 0.5 in-plane displacement). The DJ phase features a perfect stacking (0, 0 in-plane displacement) and bivalent organic cation spacers. In the ACI phase, A-site cations not only appear in the cuboctahedral cages but also alternate with the spacers in the organic layer. Many linear monoammonium cations, such as butylammonium, pentylammonium, and hexylammonium, can adopt RP structures [86]. DJ phase is usually derived from diammonium spacers, including both linear types like NH$_3$C$_m$H$_{2m}$NH$_3$ (m = 7−9) and cyclic types like 3-(aminomethyl)piperidinium (3AMP) and 1,5-naphthalene diammonium [85,87,88]. However, unlike RP and DJ phases, ACI structure can only be templated by guanidinium (GUA) currently [70].

All three phases of (quasi-)2D LHP show a conductivity in the stacking direction much lower than the in-plane directions because the organic cations act as barriers. As a result, excitons are confined in the inorganic slabs, showing a binding energy depending on the n value. Therefore, red shifts in absorption and emission spectra are observed as n increases. Although the electrical and optical properties of the three phases are similar in general, there are still some differences. DJ (quasi-)2D LHPs usually show a smaller bandgap than RP ones with the same n value due to less distortion in inorganic layers and a shorter interlayer distance [89]. In addition, DJ (quasi-)2D LHPs are more stable than RP ones [8,90]. Although the hydrophobic property of the spacers in the RP phase provides resistance to humidity, the vdWs gap in the bilayer made the devices vulnerable when being subjected to high temperature and high humidity at the same time. The hydrogen-bonding interaction in the DJ phase is stronger than the vdWs interaction in the RP phase and therefore leads to a more rigid and tight structure, providing better device stability.

For the synthesis of (quasi-)2D LHPs as a device component, a spin-coating method with a precursor is more frequently used, resulting in a multicrystalline film [70,85,91]. The precursor can be made by dissolving individual components in an organic solvent. For optoelectronic applications, the (quasi-)2D LHP layer usually consists of grains with random orientations. As the organic spacers inhibit charge transport along the stacking direction, a crystal orientation with organic layers perpendicular to the substrate is preferred in devices. Various additives, such as NH$_4$SCN, Pb(SCN)$_2$, and MACl, have been applied to the precursor to form crystals with vertically aligned organic layers [66,92,93]. It is proposed that the additives function by suppressing the formation of PbI$_2$ sol-gel which acts as nucleation sites and tends to transform into an unoriented intermediate phase [94]. Currently, grazing-incident wide-angle X-ray scattering is the most widely used tool to characterize the crystal orientation. The diffraction pattern will be a Scherrer ring if the grains are randomly oriented, and bright spots if vertically oriented (Figure 3d) [66].

In addition to random crystal orientation, the 2D LHP layer is typically composed of phases with different n values. Since the bandgap of 2D LHP decreases as n increases, the distribution of n values would cause excitons to funnel to domains with large n values [95]. Ma and coworkers designed a bifunctional molecular additive, tris(4- fluorophenyl)phosphine oxide (TFPPO), to narrow down the n-value distribution and passivate the lateral sides of 2D LHP at the same time. The n-value distribution was ascribed to the slower diffusion of organic spacers due to hydrogen bonds formed between the spacer's ammonium tails and the fluorine atoms in TFPPO [96].

A thin 2D LHP layer can also be deposited on a 3D LHP film to form a 2D/3D heterostructure by spin-coating a solution containing organic spacers on the 3D LHP film [97–99]. The resultant 2D LHP layer is typically a couple of nanometers thick and consists of phases with n = 1 or 2. Although the large bandgap and exciton binding energy of low n values are unfavorable in solar cells, the 2D LHP layers provided passivation to the surface of 3D LHP film and therefore improved the stability and open-circuit voltage. Typically, IPA is used as the solvent for organic spacers due to the good solubility it offers. Additional halide compounds, such as potassium iodide (KI) and rubidium iodide (RbI), may also be added to the solution [100]. However, IPA can also dissolve the underlying 3D LHP layer due to its strong polarity, leading to a 3D/2D mixed phase at the surface. To obtain a clear heterojunction, Yoo et al. employed chloroform (CF) as the solvent for the organic spacers to avoid damaging the underlying 3D LHP film [101]. However, only certain linear ammonium halide, for example, n-hexylammonium bromide, showed solubility high enough to be applied in the spin-coating method. To overcome the conflict in the solution process, Jang and coworkers developed a solid-phase in-plane growth method to synthesize an intact and clear 2D/3D junction with control over the thickness of the 2D layer [102]. A 2D LHP film and a 3D LHP film were deposited on two substrates separately via the single-step spin-coating method. After the two films were stacked in contact, heat and pressure were applied to induce the growth of the 2D layer on the 3D film. Finally, the substrate with 2D LHP solid precursor was detached, leaving an intact 2D layer on the 3D LHP film. The thickness of the 2D layer could be controlled by adjusting the temperature or performing the process iteratively.

3.3. 0D Lead Halide Perovskite Quantum Dots

Perovskite quantum dots (PQDs) not only inherit excellent optical and electronic properties from LHPs but also show unique advantages, such as size-tunable bandgaps, and high photoluminescence quantum yield (PLQY). PQDs are typically synthesized through a solution method. However, they may also form in inorganic glass with the aid of laser irradiation. For example, Huang et al. demonstrated the laser-induced in-situ formation of CsPbBr$_3$ quantum dot in an inorganic oxide glass which enabled 3D patterning [103]. The pattern could be erased by thermal annealing and rewritten by laser irradiation many times. Below we introduce two main solution synthesis methods for PQDs: ligand-assisted reprecipitation (LARP) and hot injection where the hot injection method is mostly employed to synthesize high-quality PQDs.

3.3.1. Ligand Assisted Reprecipitation

Generally, organic-inorganic hybrid PQDs are synthesized via the LARP method whereas the hot injection method is used more frequently for all inorganic PQDs. Schmidt et al. first reported the spherical CH$_3$NH$_3$PbBr$_3$ (MAPbBr$_3$) PQDs synthesized by a precipitation method at room temperature [104]. The PQDs showed a green emission (~525 nm) with a narrow FWHM (~25 nm) and PLQY of 20%. Later, PLQY was improved to 83% by changing the molar ratio of reactants [105]. Zhang et al. modified the precipitation method by using capping ligands (octylamine and oleic acid) to limit the growth of crystals and named it LARP (Figure 3e) [67]. A precursor with capping ligands in a good solvent (e.g., DMF) was drop-wisely added to a bad solvent (e.g., toluene). The authors concluded that octylamine controlled the kinetics of crystallization and consequently the size of PQDs, while OA prevented aggregation and therefore enhanced stability. However, organic-inorganic hybrid PQDs synthesized by the LARP method were still less stable than their all-inorganic counterpart, such as $CsPbX_3$ PQDs, due to the volatile organic component, especially when exposed to heat and moisture.

3.3.2. Hot Injection

Hot injection synthesis of PQDs was first performed by L. Protesescu et al. motivated by reports on hybrid organic-inorganic metal halide perovskite [68]. A Cs-oleate stock solution was injected into a lead halide (PbX_2) precursor at 140–200 °C to form monodispersed cubic-shape PQDs of different sizes (4–15 nm). Various halide compositions could be readily achieved by adjusting the ratio in PbX_2 precursor, such as $PbI_2:PbBr_2$ = 1:1 and $PbBr_2:PbCl_2$ = 1:1. The composition and quantum confinement effect due to sizes gave $CsPbX_3$ PQDs a tunable bandgap covering the entire visible spectral region (Figure 3f). The all-inorganic PQDs also showed a large charge carrier mobility (4500 cm^2 V^{-1} s^{-1}), high PLQY, and narrow full with half maximum (FWHM), which made them attractive in optoelectronic applications, such as solar cells, photodetectors, and LEDs [106–112]. Later, Protesescu et al. also reported a fast anion exchange in $CsPbX_3$ which provided a post-synthesis method to finely tune the composition [113]. The fast exchange was attributed to the well-known high anion conductivity in bulk halide perovskite and the highly dynamic ligand binding of the PQDs (Figure 3g) [69]. However, the easy loss of capping ligands also made it difficult to maintain PQDs intact during a purification process. Swarnkar et al. found that methyl acetate (MeOAc) was a suitable antisolvent to isolate $CsPbI_3$ PQDs while keeping the integrity [112]. Unlike their bulk counterpart which quickly transferred to an orthorhombic phase at room temperature, the densely packed PQD film was stable in a cubic α-phase for months under ambient conditions due to the large surface area and high surface energy of PQDs. Although MeOAc maintained PQDs intact during purification, there were still surface defects formed, such as iodine vacancy, while removing surface ligands. Recently, Jia et al. tackled the problem by adding tert-butyl iodide (TBI) and trioctylphosphine (TOP) in the purification process [114]. The iodine ions released from the nucleophilic substitution reaction between TBI and TOP cured the surface defects. In addition to purification methods, researchers have optimized other synthesis parameters, such as the ratio of different precursors, reaction temperatures, solvents, and capping ligands [115–120].

For PQDs, ligand exchange is necessary to fabricate efficient devices, as the long organic ligands used in synthesis significantly block the charge transfer. A two-step solid-state ligand exchange is widely performed by dipping the PQD film in methyl acetate and ethyl acetate ligand solutions subsequently [121–123]. In this way, OA/OAm capping ligands can be changed into acetic acid/FA or other short ligands to facilitate electric coupling between PQDs. However, ligand exchange may cause more defects on QD surfaces, and therefore a post-treatment with cesium salts is needed to passivate those defects [124]. Furthermore, FA cations are hygroscopic and able to enter PQDs, which results in device instability. To overcome the drawbacks of using the FA cations and enhance the device stability, other large hydrophobic cations such as phenylethylammonium (PEA) and GUA have been explored [125,126].

4. 2D Materials for Optoelectronic Applications

2D materials, such as graphene, TMDCs, and BP, have attracted tremendous research interest in the field of optoelectronics due to their facile processing techniques and unique properties arising from vdWs structures [33]. For instance, graphene shows ultrafast carrier dynamics and a broad absorption band. TMDCs are favored for their direct bandgap, high light absorption, natural abundance, and excellent chemical stability. Besides, their mechanical flexibility and durability allow the formation of high-quality interfaces with a low density of charge traps [127]. BP has high carrier mobility and a moderated bandgap of around 0.3 eV in its single-layer form, which enables reduced dark current and low noise in photodetection [128]. In this part, we present the recent development of 2D material-based optoelectronics.

4.1. 2D Material Solar Cells

One of the MoS_2-based solar cells was proposed by Wi et al. [127]. In the work, the author treated the surface of the MoS_2 layer with CHF_3 plasma to form a p-n junction in the solar cell (Figure 4a). As illustrated in Figure 4b, the built-in potential between the plasma-treated MoS_2 and untreated MoS_2 layer could enhance the separation and collection of photogenerated carriers, resulting in improved device performance. The device showed a high current density of 20.9 mA/cm^2 and PCE of 2.8%. WSe_2 has also shown its potential as a solar cell material in the work reported by McVay et al. [129]. The WSe_2 layer was passivated with Al_2O_3 and the solar cell demonstrated remarkable photocurrent enhancement (Figure 4c). The passivation layer not only reduced surface traps but also induced n-type surface doping and band bending at the interface, increasing the active area for photocurrent extraction. The fabricated device exhibited a V_{oc} of 380 mV and J_{sc} of 10.7 mA/cm^2. In addition, Nazif et al. proposed a flexible solar cell consisting of the WS_2 absorbing layer, graphene top contact, and MoO_x coating (Figure 4d) [130]. The transparent graphene top contacts greatly reduced Fermi-level pinning at the interface of TMDC and metal contacts. Additionally, MoO_x served as effective passivation and anti-reflection layer, leading to increased J_{sc}. As a result of these strategies, the device exhibited remarkable performance enhancement, with a PCE of 5.1% and specific power of 4.4 W/g in comparison with that of the previously reported counterpart (PCE ~0.7%, specific power ~0.04 W/g). The device performance of solar cells reviewed in Section 4.1 is summarized in Table 1.

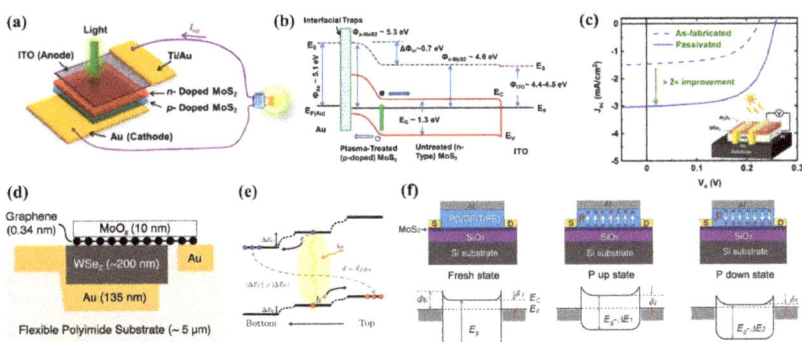

Figure 4. (a) Schematic of the plasma-treated MoS_2 solar cell. (b) Energy band level of the plasma-treated MoS_2 solar cell. (c) The performance improvement of the solar cell passivated by Al_2O_3 under a 3400 K black body source of 30 mW/cm^2. (d) Cross-sectional image of the MoO_x/graphene/WSe_2 solar cell. (e) Energy band diagram of the strain-gradient WS_2 film. (f) The schematic and energy band diagram of the MoS_2 device under different polarization states. (a,b) Reprinted with permission from Ref. [127]. Copyright 2014, ACS. (c) Reprinted with permission from Ref. [129]. Copyright 2020, ACS. (d) Reprinted with permission from Ref. [130]. Copyright 2021, Springer Nature. (f) Reprinted with permission from Ref. [131]. Copyright 2015, WILEY. (e) Reprinted with permission from Ref. [132]. Copyright 2021, Wiley.

Table 1. Summarized device performance of solar cells in Section 4.1.

Active Materials	Voc (V)	Jsc (mA/cm^2)	FF (%)	PCE (%)	Ref.
MoS$_2$	0.28	20.9	47	2.8	[127]
WSe$_2$	0.38	10.7	44	1.6	[129]
WS$_2$	0.476	17.3	61.7	5.1	[130]

4.2. 2D Material Photodetectors

4.2.1. Graphene

Ever since the successful exfoliation, graphene has been widely applied to photodetector applications. Here, we summarize significant works based on graphene. First, a graphene photodetector was integrated into a silicon bus waveguide, demonstrating a chip-integrated photodetector technology based on graphene [133]. The coupling strategy in the work greatly improved the light absorption capability of the graphene layer over a broadband spectrum. As a result, the photodetector achieved maximum responsivity of over 0.1 A/W and uniform photoresponse ranging from 1450 to 1590 nm. Second, Kim et al. fabricated a photodetector based on an all-graphene p-n junction [134]. Attributed to the carrier multiplication induced by impact ionization and high photoconductive gain in the graphene layers, the device demonstrated high detectivity (~10^{12} Jones) and photoresponsivity (0.4–1.0 A/W) from ultraviolet to near-infrared region. However, due to the emergence of TMDCs, graphene was substituted with TMDCs for photodetector applications.

In recent years, waveguide-integrated graphene photodetectors have been proposed for datacom applications. Schuler et al. used a silicon photonic waveguide to guide the light in a confined graphene area to facilitate light-matter interaction [135]. The device showed a responsivity of 0.17 A/W. Following this work, Ma et al. demonstrated a graphene photodetector coupled with a silicon waveguide, and arrayed metallic structures were introduced into the device structure to improve the responsivity by exciting surface plasmon polaritons [136]. As a result, the device exhibited an external efficiency of 0.5 A/W with an ultrahigh-frequency response over 110 GHz. Similarly, Alaloul et al. reported a CMOS-compatible graphene photodetector with a plasmon-enhanced responsivity of 1.4 A/W and broad bandwidth exceeding 100 GHz [137].

4.2.2. Transition Metal Dichalcogenides

To date, various TMDCs have been applied for the photodetectors, such as WS$_2$, MoTe$_2$, and MoS$_2$ [138]. For example, WS$_2$ TMDC has attracted tremendous interest as the photodetector material attributed to its availability in both n- and p-type. WS$_2$ is intrinsically an n-type semiconductor, but through doping p-type WS$_2$ can be readily attained. Kim et al. introduced a gradient-strained WS$_2$ film, which was controlled by the sputtering time of W, for the photodetector application [132]. Due to the band edge variation resulting from the gradient strain, a type II homojunction was formed in the multilayer film which effectively reduced nonradiative recombination (Figure 4e). The photodetector demonstrated high photoresponsivity under white light irradiation with a low driven voltage of 0.1 V.

Similarly, Mo-based TMDCs, such as MoS$_2$ and MoTe$_2$, are intrinsically an n-type TMDC with availability in a p-type as well. Particularly, MoTe$_2$ has a suitable bandgap for both visible and near-infrared light photodetection [139,140]. Therefore, a broadband photodetector was introduced by employing the MoTe$_2$ TMDC [141]. The device exhibited high detectivity of 3.1×10^9 and 1.3×10^9 Jones for 637 and 1060 nm light, respectively. The authors attributed the high device performance to the photogating effect. The photogenerated holes were localized within the trap states, leading to high photoconductive gain. In addition, Wang et al. developed a MoS$_2$-based photodetector with P(VDF-TrFE) as the ferroelectric gate [131]. The ferroelectric material provided an ultrahigh electrostatic field of about 10^9 V/m in the semiconducting channel and helped the device maintain a depleted state, significantly suppressing the dark current (Figure 4f). As a result, the device showed broadband photodetection from 0.85–1.55 µm, as well as high detectivity of 2.2×10^{12} Jones and responsivity of 2570 A/W.

4.2.3. Black Phosphorus

BP, a layered semiconducting material similar in appearance to graphite, has risen as a new family member of 2D materials since its first successful exfoliation in 2014. BP has brought new concepts and applications to the field of optoelectronics due to its unique structure and properties. Particularly, BP has the infrared bandgap which offers great potential for IR detection [142]. For example, a mid-infrared photodetector based on BP was demonstrated by Guo et al. [128]. The photodetector exhibited a wide range of operating wavelengths from 532 nm to 3.39 µm. The device showed responsivity of 82 A/W at 3.39 µm and was capable of detecting picowatt light owing to its high photoconductive gain and low dark current. Moreover, the photodetector had effective photoresponse in the kilohertz frequency range due to the carrier dynamics of BP with a moderate bandgap. In addition, Jalaei et al. reported a Se-doped BP photodetector with high responsivity ranging from 3 to 5 µm [143]. Se-doped BP showed smaller bandgap as well as higher absorption coefficient in the mid-infrared region compared to the pristine BP. As a result, the device exhibited responsivity of 0.75 µA/W at 4 µm with the light intensity of 0.0001 W/cm^2.

Despite the great progress in optoelectronics using 2D materials as active layers, the majority of devices exhibited unsatisfactory performance mainly due to the poor light absorption limited by their atomic-scale thickness. To resolve this issue, heterostructures were introduced and this structure improved the light absorption capacity, which will be discussed in Part 6 [144]. The device performance of photodetectors reviewed in Section 4.2 is summarized in Table 2.

Table 2. Summarized device performance of photodetectors in Section 4.2.

Active Materials	Response Range	R (A/W)	D* (Jones)	τ_{rise}/τ_{fall}	Ref.
Graphene	1450–1590 nm	0.108	-	-	[133]
Graphene	UV-NIR	~1	10^{12}	-	[134]
Graphene	-	0.17	-	-	[135]
Graphene	1480–1620 nm	0.5	-	-	[136]
Graphene	-	1.4	-	50 µs/-	[137]
WS$_2$	440–800 nm	30	-	-	[132]
MoTe$_2$	600–1550 nm	0.05	3.1×10^9	1.6 ms/1.3 ms	[141]
MoS$_2$	500–1550 nm	2570	2.2×10^{12}	1.8 ms/2 ms	[131]
BP	0.532–3.39 µm	82	-	-	[128]
BP	MIR	7.5×10^{-7}	-	-	[143]

5. Perovskite Optoelectronic Applications

5.1. Lead Halide Perovskite Solar Cells

5.1.1. 3D Lead Halide Perovskite

In two decades, 3D LHP solar cells have made rapid progress and are almost catching up with the PCE of Si-based types. The record PCE of 3D perovskite in a solar cell is 25.6% while a Si-based cell is 26.1%, making perovskite an attractive material for solar cell applications. The first report on 3D LHP solar cells dates back to 2009 by Kojima et al. The device exhibited a PCE of 3.8% using MAPbI$_3$ as the active layer. [26] Following this work, in 2012, Kim et al. reported a high-efficiency solar cell with a PCE of 9.7% based on the structure of mesoporous TiO$_2$/MAPbI$_3$/spiro-OMeTAD [145]. In the same year, Lee et al. employed mesoporous alumina as an insulating scaffold for the MAPbI$_2$Cl perovskite and achieved a solar cell with a PCE of 10.9% and low energy losses [146]. Due to high PCEs achieved in a short period, these two findings have attracted intense research interest in perovskite solar cells and greatly facilitated the development of this area.

In 2013, a two-step sequential deposition method was introduced for the formation of MAPbI$_3$ perovskite layer in solar cells, which opened a new route for high-performance solution-processed perovskite solar cells [147]. PbI$_2$ dissolved in DMF solution was spin-coated on the TiO$_2$ underlayer, and then the substrate was dipped into the MAI solution. This led to PbI$_2$ being transformed into the 3D LHP with a greatly improved film morphology compared to the conventional one-step deposition approach. Attributed to the improved morphology, the device exhibited a high PCE of 15%. Similarly, Yang et al. deposited a PbI$_2$ (DMSO) complex in a DMF solution on the TiO$_2$ layer, followed by the spin-coating of the FAI solution [148]. FAPbI$_3$ crystallization occurred through the intramolecular exchange process (IEP) of DMSO with FAI. According to SEM and XRD results, the perovskite film by the IEP method showed a smoother surface, larger grain size, and better (111)-preferred crystal orientation in comparison with its counterpart by the conventional method. Benefiting from this technique, over 20% of PCE was achieved in 3D perovskite solar cells.

In 2015, Bi et al. reported high-quality MAPbI$_3$ perovskite layers grown on non-wetting HTLs with a thickness aspect ratio of 2.3–7.9 [149]. Non-wetting HTLs reduced the surface tension dragging force compared to wetting HTLs, which increased the grain boundary mobility, yielding the growth of larger perovskite grains with improved crystallinity. As a result, the charge recombination was suppressed and the fabricated device achieved a high PCE of 18.3%. Following this work, Serpetzoglou et al. compared the performance of MAPbI$_3$ solar cells utilizing the hydrophilic PEDOT: PSS and the non-wetting PTAA as the HTLs, respectively [150]. It was found that the non-wetting PTAA layer with a smooth surface favored the formation of uniform and larger perovskite grains. Additionally, faster relaxation times and slower bimolecular recombination rates were observed for PTAA-based devices. Owing to the merits provided by the PTAA HTL layer, the PCE of the solar cell showed remarkable enhancement from 12.6% to 15.67%.

Laser-assisted crystallization technique has also been proved to be effective to elevate the device's efficiency. For instance, Li et al. applied a laser irradiation approach to induce rapid crystallization and obtained homogeneous, pinhole-free perovskite layers [151]. Owing to the gradient distribution of laser irradiation, a proper amount of PbI$_2$ was formed on the perovskite surface, which led to self-passivation and suppressed surface states. The optimal solar cell exhibited a high PCE of 17.8%. Jeon et al. used low energy NIR laser beam to improve the crystallization and film morphology of MAPbI$_3$ perovskites [152]. The irradiation process helped with the perovskite phase transformation from a poor crystalline intermediate phase to a highly crystalline phase, contributing to devices with PCE of 11.3% and 8% on glass and flexible substrates, respectively. Konidakis et al. also demonstrated the preferable grain size and accelerated charge carrier extraction in perovskites by the laser-assisted method [153].

In 2019, the new world record PCE of 23.3% was reported by using a double-layered halide architecture (DHA) and Poly(3-hexylthiophene) (P3HT) as the HTL [154]. An ultrathin wide-bandgap halide (WBH) was introduced between the narrow-bandgap halide (NBH) and the HTL, which reduced carrier recombination at the perovskite/P3HT interface. The authors indicated that the alkyl chain in the WBH layer interacted with the P3HT and induced a self-assembled P3HT layer with suppressed trap states. Recently, further improvement in PCE was achieved and updated the world record for PCE of 25.6%. The author utilized the pseudo-halide anion formate (HCOO$^-$) to reduce the halide vacancies at the boundaries of perovskite films [27]. It was claimed that HCOO$^-$ anions passivated I$^-$ vacancies by interacting with undercoordinated Pb^{2+}, effectively reducing non-radiative recombination.

5.1.2. Quasi-2D Lead Halide Perovskite

Despite the high efficiency of 3D LHP solar cells, their long-term stability against oxygen and moisture has largely limited the commercialization of perovskite solar cells. Quasi-2D perovskites have shown great potential to enhance device stability because their hydrophobic organic ligands can serve as an encapsulation layer [155]. In recent years, various strategies have been utilized to improve the performance of quasi-2D LHP solar cells. In general, quasi-2D perovskite crystals tend to have random orientations where the insulating cations hinder efficient charge carrier transport through the vertical direction. As a result, quasi-2D perovskite solar cells show lower photocurrent compared to their counterparts, such as 3D and 0D perovskite solar cells [156]. Therefore, it is crucial to induce preferential vertical crystal orientation in quasi-2D perovskite. To solve this issue, Zhang et al. introduced NH$_4$SCN as an additive in (PEA)$_2$(MA)$_4$Pb$_5$I$_{16}$-based solar cells [92]. The improved vertical orientation of perovskite was confirmed by the XRD results and the detailed mechanism is shown in Figure 5a. Attributed to the enhanced carrier transport through the vertical direction, the J$_{SC}$ was increased from 0.93 mA/cm^2 without additive to 15.01 mA/cm^2 with the optimal amount of additive, and the best device exhibited a high PCE of 11.01% (Figure 5b). In addition, the phase distribution of different n values in quasi-2D perovskite is also critical to the device's performance. Shao et al. developed a vacuum-assisted method to anneal PEA$_2$MA$_4$Pb$_5$I$_{16}$ perovskite [157]. On the one hand, the vacuum-assisted method resulted in a smooth and compact film with better morphology owing to its slower crystallization speed. On the other hand, this method significantly reduced the amount of n = 2 phases located at the bottom of the perovskite film, resulting in the promoted charge carrier transport and reduced recombination. As a result, the device by the vacuum-assisted method exhibited remarkable PCE enhancement from 3.65% to 14.14%. In 2018, the new record was achieved with a PCE of 18.2% using 3-bromobenzylammonium iodide (3BBAI)-based quasi-2D LHP [158]. In 3BBAI-based perovskite films, large n phases were located at the top of the film while small n components were at the bottom. The highly-crystalized large n components facilitated photon capture and carrier separation because of small bandgap and exciton binding energy whereas the vertically-orientated small n components were beneficial for charge transport. The device retained 82% of its original efficiency after 2400 h of storage under a relative humidity of 40% without encapsulation, demonstrating its excellent moisture stability.

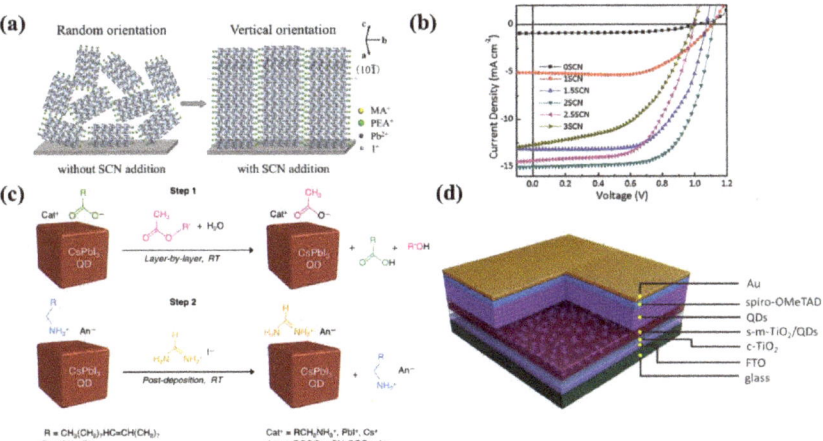

Figure 5. (a) Schematic illustration of inducing vertical crystal orientation in 2D perovskite. (b) The J–V curves for (PEA)$_2$(MA)$_4$Pb$_5$I$_{16}$-based devices with different amounts of NH$_4$SCN. (c) A two-step solid-state QD ligand-exchange procedure of CsPbI$_3$ PQDs. (d) Schematic of the PQDSC based on CsPbI$_3$ QDs and Cs-treated m-TiO$_2$ layers. (**a,b**) Reprinted with permission from Ref. [92]. Copyright 2018, Wiley. (**c**) Reprinted with permission from Ref. [122]. Copyright 2018, ACS. (**d**) Reprinted with permission from Ref. [159]. Copyright 2020, ACS.

5.1.3. 0D Lead Halide Perovskite Quantum Dots

0D PQD solar cells have also demonstrated stable device performance compared to their 3D counterparts owing to the large surface area and high surface energy of PQDs. In addition, most trap states of PQDs are located in the conduction and valence bands attributed to the low defect formation energy. Therefore, PQDs exhibit defect-tolerant characteristics, which is favorable to optoelectronic applications [160]. However, the organic insulating ligands on the dot surface hinder effective carrier transport within the PQD film. Therefore, ligand exchange is required to substitute long-chain ligands with shorter ones in PQD solar cells. A two-step solid-state ligand-exchange method was introduced by Wheeler et al. to remove native oleate and oleylammonium ligands of $CsPbI_3$ PQDs (Figure 5c) [122]. Through the application of methyl acetate and formamidinium iodide in the first and second step, respectively, the PQD solid film showed improved charge transport and consequently, a PCE of 12% was achieved.

The engineering of charge transport layers is also critical for fabricating high-performance PQD solar cells. Chen et al. applied Cs-ion-containing methyl acetate solution to the m-TiO_2 electron transport layer and improved interfacial properties between m-TiO_2 and PQDs (Figure 5d) [159]. The Cs^+ ions promoted the incorporation of PQDs into the m-TiO_2 layer and passivated the PQD surface, resulting in an enhanced electron injection rate. Additionally, the authors utilized an ethanol-environment smoothing approach for reducing the surface roughness of the m-TiO_2 film. Attributed to those approaches, the optimized device exhibited a high PCE of 14.3%.

Other strategies such as site doping have also been explored. Hao et al. fabricated $Cs_{1-x}FA_xPbI_3$ PQD films with reduced defect density with the aid of OA ligands [121]. The authors demonstrated that the surface ligands were crucial for the formation of A-site vacancies and OA ligands greatly facilitated cation cross-exchange. The optimized device based on $Cs_{0.5}FA_{0.5}PbI_3$ PQDs showed an excellent PCE of 16.6% which is the highest record to date. Attributed to suppressed phase segregation in PQDs, the PQD-based devices showed superior long-term photostability compared to their 3D counterparts. The device performance of solar cells reviewed in Section 5.1 is summarized in Table 3.

Table 3. Summarized device performance for solar cells in Section 5.1.

Active Materials	Voc (V)	Jsc (mA/cm^2)	FF (%)	PCE (%)	Ref.
$MAPbI_3$	0.61	11	57	3.81	[26]
$MAPbI_3$	0.888	17.6	62	9.7	[145]
$MAPbI_2Cl$	0.98	17.8	63	10.9	[146]
$MAPbI_3$	0.992	17.1	73	12.9	[147]
$FAPbI3$	1.06	24.7	77.5	20.2	[148]
$MAPbI_3$	1.07	22	76.8	18.3	[149]
$MAPbI_3$	1.01	20.24	76.67	15.67	[150]
$MAPbI_3$	1.146	22.82	68	17.8	[151]
$MAPbI_3$	0.91	14.5	80	11.3	[152]
$(FAPbI_3)_{0.95}(MAPbBr_3)_{0.05}$	1.152	24.88	81.4	23.3	[154]
$FAPbI_3$	1.189	26.35	81.7	25.6	[27]
$(PEA)_2(MA)_4Pb_5I_{16}$	1.11	15.01	67	11.01	[92]
$(PEA)_2(MA)_4Pb_5I_{16}$	1.1	17.52	73	14.14	[157]
3BBAI-based quasi-2D	1.23	18.22	81.2	18.2	[158]
$CsPbI_3$ QDs	1.2	-	-	12	[122]
$CsPbI_3$ QDs	1.06	17.77	75.8	14.32	[159]
$Cs_{0.5}FA_{0.5}PbI_3$ QDs	1.17	18.3	78.3	16.6	[121]

5.2. Perovskite Photodetectors

5.2.1. 3D Lead Halide Perovskite

Attributed to their excellent properties, perovskites have been actively applied to photodetection technology as well. For example, a broadband photodetector with a range of 240 to 750 nm was achieved by utilizing Cs-doped $FAPbI_3$ perovskite [29]. The Cs-doping improved film quality and the favorable Schottky junction was formed between the Au and the perovskite layer (Figure 6a). As a result, the device showed a rapid response speed with a rise and fall time of 45 ns and 91 ns, respectively. In addition, highly-sensitive photodetectors were reported based on $CsPbI_xBr_{3-x}$ perovskites [161]. In the work, the authors treated the surface of PTAA with amphiphilic PEIE to improve its wettability, leading to enhanced film coverage and quality of the above perovskite layer. The best device based on CsPbIBr$_2$ exhibited a high detectivity of 9.7×10^{12} Jones and ultrafast response speed of 20 ns. The device remained 87% of its original photoresponsivity after more than 2000 h of storage in ambient air, which demonstrated its impressive environmental stability.

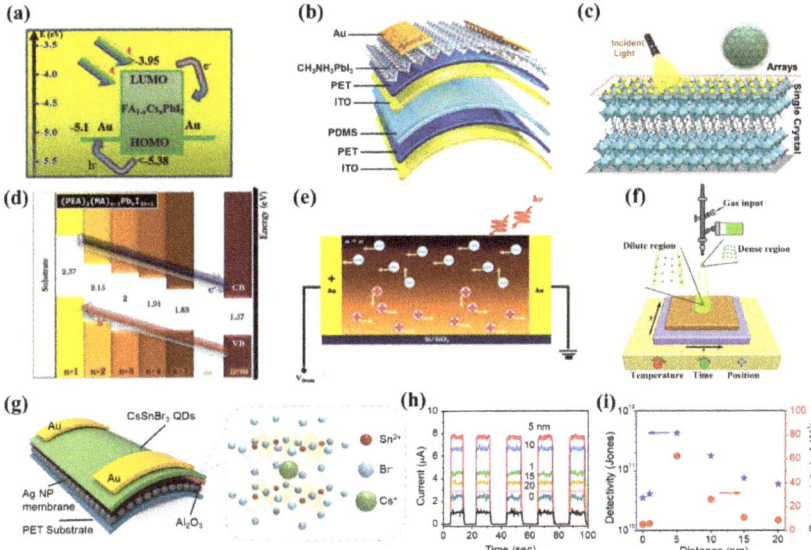

Figure 6. Schematic illustration of (**a**) the energy level of Cs-doped FAPbI$_3$ perovskite photodetector under incidental light, (**b**) the MAPbI$_3$-based flexible photodetector, (**c**) the array photodetector based on (PA)$_2$(G)Pb$_2$I$_7$. Schematic of (**d**) the energy band diagram and (**e**) charge transport in the gradient 2D/3D perovskite films. (**f**) Schematic illustration of the spray-coating deposition method. (**g**) Schematic of the PET/Ag NP/Al$_2$O$_3$/CsSnBr$_3$ QD photodetector. (**h**) Photocurrent of the photodetectors with different thicknesses of the Al$_2$O$_3$ layers. (**i**) The detectivity and responsivity of photodetectors with different thicknesses of the Al$_2$O$_3$ layers. (**a**) Reprinted with permission from Ref. [29]. Copyright 2017, WILEY. (**b**) Reprinted with permission from Ref. [162]. Copyright 2018, WILEY. (**c**) Reprinted with permission from Ref. [163]. Copyright 2019, WILEY. (**d**,**e**) Reprinted with permission from Ref. [164]. Copyright 2020, WILEY. (**f**) Reprinted with permission from Ref. [28]. Copyright 2018, ACS. (**g**–**i**) Reprinted with permission from Ref. [165]. Copyright 2021, ACS.

Furthermore, a self-powered photodetector was realized using MAPbI$_3$ as the light-harvesting layer on a flexible substrate (Figure 6b) [162]. Drop-by-drop solvent engineering of toluene was applied during spin-coating the perovskite layer to slow down the crystallization process, resulting in improved film morphology. As a result, the fabricated photodetector exhibited excellent detectivity of 1.22×10^{13} Jones. Additionally, the flexible device was capable of harvesting omnidirectional light owing to its transparent polymer substrate. Remarkably, the device exhibited excellent physical stability even after bending 1000 times.

5.2.2. Quasi-2D Lead Halide Perovskite

In the case of quasi-2D LHP, a new perovskite framework employing G cation was reported for an array photodetector application (Figure 6c) which sheds light on a new branch of 2D hybrid perovskite [163]. In the work, the authors used (PA)$_2$(G)Pb$_2$I$_7$ perovskite crystals (where PA is n-pentylaminium and G is guanidinium), and achieved reduced dark current by the high-quality single crystals as well as the amplified photocurrent by the array structure. As a result, the photodetector exhibited superior performance such as high detectivity (6.3×10^{12} Jones) and ultrafast response speed (3.1 ns). 2D/3D gradient $(PEA)_2(MA)_{n-1}Pb_nI_{3n+1}$ perovskite structure was employed to fabricate high-performance perovskite photodetectors [164]. As illustrated in Figure 6d, the favorable energy band alignment along the vertical heterojunctions facilitated the separation of electron-hole pairs. Additionally, carrier recirculation in the gradient perovskite films enabled a long carrier lifetime (Figure 6e). As a result, a highly sensitive photodetector was obtained with a responsivity of 149 A/W and a detectivity of 2×10^{12} Jones.

In addition to LHP, a lead-free green photodetector was realized by integrating a 2D $(PEA)_2SnI_4$ perovskite microsheet using a ternary-solvent method [166]. Under the light illumination of 195.8 µW/cm^2, the device showed photoresponsivity of 3.29×10^3 A/W which is higher than its counterpart based on $(PEA)_2SnI_4$ polycrystalline film [167]. The high detectivity of 2.06×10^{11} Jones and photoconductive gain of 8.68×10^3 was also demonstrated. This work opened up great possibilities for the high-performance photodetector for green optoelectronics.

5.2.3. 0D Lead Halide Perovskite Quantum Dots

To improve the air stability of perovskite-based photodetectors, PQDs have been integrated into photodetectors. For example, Zhang et al. fabricated air-stable α-CsPbI$_3$ PQD film which exhibited high stability in material properties after exposure at a relative humidity of 30% for 60 days [168]. The authors modified the surface of α-CsPbI$_3$ QDs utilizing an up-conversion material (NaYF$_4$:Yb,Er QDs) to extend its absorption region and achieved broadband photodetection from 260 nm to 1100 nm. Additionally, a high on/off ratio of 10^4, a responsivity of 1.5 A/W, and a short response time (less than 5ms) were also demonstrated. In spite of their high air stability, the yield of PQDs is relatively low compared to their 3D and quasi-2D counterparts. Furthermore, a large volume of materials is wasted during a conventional spin-coating method, limiting the scalability of PQD-based photodetectors. To resolve this issue, a spray-coating approach was developed to deposit the CsPbBr$_3$ QD layer in a photodiode (Figure 6f) [28]. This spray technique enabled a high material utilization rate of 32% compared to that of the spin-coating method (only 1%). Additionally, the authors precisely adjusted the substrate temperature and spray time to obtain the crack-free PQD film and demonstrated a large-area photodiode of 10×10 cm^2 with a high detectivity of 1×10^{14} Jones.

In addition to LHP-based PQDs, lead-free PQDs have also been actively researched to realize an environmentally-friendly photodetector. Ma et al. synthesized lead-free CsSnBr$_3$ QDs and introduced a plasmonic/dielectric nanostructure to optimize the performance of a flexible photodetector (Figure 6g) [165]. Plasmonic Ag nanoparticles (NPs) were applied to improve the light absorption capability of the CsSnBr$_3$ PQDs through localized surface plasmon resonance. A dielectric Al$_2$O$_3$ spacer layer was inserted between the PQD layer and the Ag membrane to reduce the surface energy quenching. As shown in Figure 6h, with the optimized thickness of the Al$_2$O$_3$ layer, the photocurrent showed a remarkable enhancement which was 7.5 times higher than that of the device without the dielectric layer. The self-assembled Ag NPs membrane could release some bending tension, which enabled the flexible device with good mechanical durability. The photodetector showed the maximum detectivity of 4.27×10^{11} Jones and responsivity of 62.3 mA/W (Figure 6i), which are among the highest values for Sn-based perovskite photodetectors.

Perovskites have great potential for future generation optoelectronic applications due to their fascinating material properties. Nevertheless, some prominent issues regarding the device structures and inherent material defects hampered their further development towards commercialization. First, there is an urgent need for replacing the commonly used carrier transport materials. On the one hand, most perovskite optoelectronic devices utilized PEDOT: PSS as the hole transport layer, whose acidity and hygroscopicity would severely deteriorate the device's durability [169]. On the other hand, as the popular electron transport material, TiO$_2$ suffered from a high density of trap states and the required high-temperature sintering (400–500 °C) process [170]. In addition, the devices generally showed poor stability and large recombination loss due to the sensitivity and a great number of defects of perovskites. The introduction of passivation and encapsulation layers is expected to solve these problems. The device performance of photodetectors reviewed in Section 5.2 is summarized in Table 4.

Table 4. Summarized device performance of photodetectors in Section 5.2.

Active Materials	Response Range (nm)	R (A/W)	D* (Jones)	τ_{rise}/τ_{fall}	Ref.
$FA_{0.85}Cs_{0.15}PbI_3$	240–750	5.7	2.7×10^{13}	45 ns/91 ns	[29]
$CsPbIBr_2$	400–580	0.28	9.7×10^{12}	20 ns/20ns	[161]
$MAPbI_3$	-	0.418	1.22×10^{13}	-	[162]
$(PA)_2(G)Pb_2I_7$	420–700	~47	6.3×10^{12}	0.94 ns/2.18 ns	[163]
$(PEA)_2(MA)_{n-1}Pb_nI_{3n+1}$	420–760	149	2×10^{12}	69 ms/103 ms	[164]
$(PEA)_2SnI_4$	-	3.29×10^3	2.06×10^{11}	0.37 s/3.05 s	[166]
$CsPbI_3$ QDs	260–1100	1.5	-	<5 ms/<5 ms	[168]
$CsPbBr_3$ QDs	-	3	10^{14}	-	[28]
$CsSnBr_3$ QDs	300–630	0.0623	4.27×10^{11}	50 ms/51 ms	[165]

6. 2D Materials/Perovskites Interface Engineering

As we discussed above, 2D materials and perovskites have achieved great success during the past decade in the field of optoelectronics. However, challenges still remain and there is high demand for further improving the performance of the devices. For instance, the performance of 2D materials-based optoelectronics is generally limited by weak light absorption caused by their atomic-scale thicknesses, whereas perovskites-based devices suffer from severe stability issues against oxygen and moisture [33]. In recent years, the concept of 2D material/perovskite heterostructure has been proposed. Devices based on this novel structure exhibited improved performance and stability by taking advantage of the outstanding optical properties of perovskites and the fascinating electrical and optical properties of 2D materials in addition to an encapsulation effect. Furthermore, there are large families of 2D materials and perovskites, opening up great possibilities for engineering the heterostructures [33]. In this part, we summarize recent achievements in solar cells and photodetectors based on 2D material/perovskite heterostructure.

6.1. 2D Material/Perovskite Heterostructure for Solar Cells

2D materials have excellent material properties, such as low trap density and high carrier mobility. Therefore, they have been considered promising candidates for charge transport layers in perovskite solar cells [171].

6.1.1. Graphene Derivative/Perovskite Heterostructure

Derivatives of graphene have been widely utilized in perovskite solar cells to improve device performance and stability. For example, graphene oxide (GO) was introduced as a hole transport layer (HTL) in $MAPbI_{3-x}Cl_x$-based solar cells by Wang et al. [172]. The pristine GO was treated with ammonia, which helped with improved film coverage and crystallinity of the perovskite layer. It was reported that the better wettability of ammonia-treated GO to DMF precursors, as well as the strong Pb–N coordination bond between perovskites and a-GO, enhanced the crystallization, yielding a PCE of 14.14%. However, the application of GO in high-performance perovskite solar cells is largely restricted because of its insulating characteristics and a high degree of surface oxygen [173]. To solve this issue, rGO was employed as the HTL layer to the perovskite solar cell application as the rGO has low surface oxygen content and an intrinsic passivation effect towards water and oxygen [174]. Jokar et al. demonstrated that the reduction of oxygen atoms decreased charge recombination caused by the localized holes at the rGO/perovskite interface, contributing to the elevated PCE of 16.4% [175]. Similarly, Chen et al. substituted acid and hydrophilic PEDOT: PSS with covalently sulfated graphene oxide (oxo-G1) and employed it as the HTL layer [169]. Oxo-G1 reduced the speed of water ingress into the device and improved its stability in ambient air. As a result, the rGO/MAPbI$_3$ heterostructure device demonstrated enhanced PCE and much slower degradation over time compared to devices with PEDOT: PSS as HTL. Another approach was introduced by Agresti et al. The authors inserted lithium neutralized graphene oxide (GO-Li) as the electron transport layer (ETL) in MAPbI$_3$ perovskite solar cells (Figure 7b) [176]. The GO-Li formed a favorable band alignment with titanium dioxide (TiO$_2$), which significantly facilitated electron extraction from the perovskite to the mesoporous TiO$_2$ (m-TiO$_2$) layer. Compared to the reference sample without GO-Li, the fabricated device showed increased J$_{sc}$ (+10.5%) and FF (+7.5%) as well as reduced hysteresis. Additionally, GO-Li passivated oxygen vacancies of m-TiO$_2$, resulting in enhanced stability under illumination.

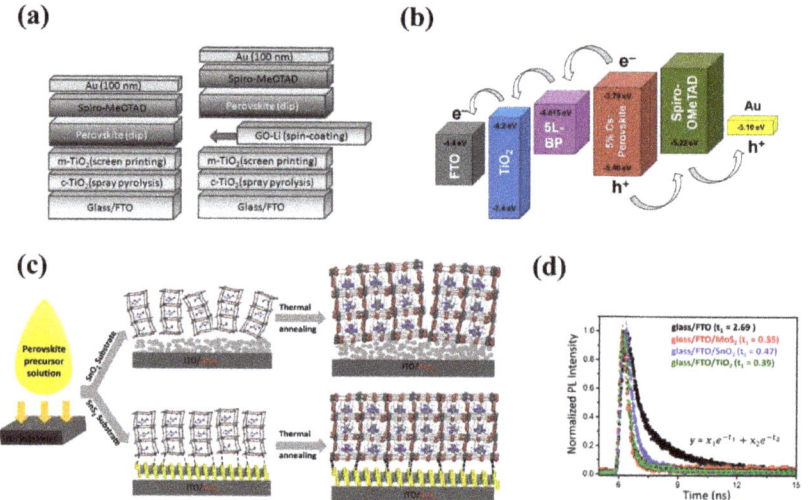

Figure 7. (a) Schematic structure of the reference device and GO-Li device. (b) Schematic illustration of energy diagram and carrier transfer in the perovskite solar cell. (c) The schematic illustration of the formation of perovskite films on SnO$_2$ and SnS$_2$ ETLs. (d) The TRPL results of perovskite films with different ETLs. (a) Reprinted with permission from Ref. [176]. Copyright 2016, WILEY. (b) Reprinted with permission from Ref. [177]. Copyright 2019, WILEY. (c) Reprinted with permission from Ref. [178]. Copyright 2018, WILEY. (d) Reprinted with permission from Ref. [170]. Copyright 2019, RSC.

6.1.2. TMDC/Perovskite Heterostructure

Similar to the derivatives of graphene layers, TMDC has been widely employed as a charge transport and/or passivation layer as well. Singh et al. synthesized an MoS$_2$ layer by a facile microwave-assisted method as an ETL in MAPbI$_3$ perovskite solar cells for the first time [170]. As the MoS$_2$ has high electron mobility and a low density of trap states, a perovskite film with the MoS$_2$ layer exhibited reduced nonradiative recombination compared to the perovskite solar cell with TiO$_2$. Particularly, as the TRPL results show in Figure 7c, the perovskite film with the MoS$_2$ layer as a quencher showed the shortest recombination lifetime, indicating the efficient charge transfer of the MAPbI$_3$/MoS$_2$ heterostructure. The as-obtained solar cell exhibited a PCE of 13.14%, which was comparable to that of TiO$_2$- and SnO$_2$-based devices. Similarly, an SnS$_2$ layer was employed as the ETL for the FA$_{0.75}$MA$_{0.15}$Cs$_{0.1}$PbI$_{2.65}$Br$_{0.35}$ perovskite solar cell where the SnS$_2$ layer was prepared by a self-assembled stacking deposition method [178]. The SnS$_2$ nanosheet induced heterogeneous nucleation of the perovskite and enabled uniform film with homogeneous grain size. The interaction between Pb and S atoms at the SnS$_2$/perovskite interface effectively passivated defect states and improved charge extraction (Figure 7d). The fabricated device showed a higher PCE of 20.12% in comparison with the conventional SnO$_2$-based device (17.72%).

The TMDC/perovskite heterostructure was also widely applied to the HTL. For example, Kim et al. fabricated MAPbI$_{3-x}$Cl$_x$ solar cells utilizing MoS$_2$ and WS$_2$ as the HTL [179]. Attributed to the suitable work functions that MoS$_2$ and WS$_2$ have, the MoS$_2$- and WS$_2$-based perovskite solar cells exhibited a PCE of 9.53% and 8.02%, which are comparable to PEDOT: PSS-based devices (9.93%). Similarly, MoSe$_2$ was employed as HTL to form the MoSe$_2$/MAPbI$_3$ heterostructure in the perovskite solar cell, and a PCE of 8.23% was attained [180]. The authors discovered that increasing the annealing temperature of MoSe$_2$ from 500, 600 to 700 °C had a positive effect on the perovskite crystallization, resulting in improved device performance.

6.1.3. Other 2D Material/Perovskite Heterostructure

Finally, other 2D materials have also been widely employed in perovskite solar cells. A phosphorene, an isolated single layer of BP, is one of the promising candidates for the heterostructure because it has a high carrier mobility of over 1000 cm^2 V^{-1} s^{-1} and formed good energy band alignment in the device structure [177]. By using a simple vortex fluidic mediated exfoliation method, the phosphorene nanosheet was attained and applied to the perovskite solar cell as an ETL (Figure 7e). The phosphorene nanosheet effectively promoted electron transfer, and therefore the average PCE of the devices improved from 14.32% to 16.53%. Recently, a Ti$_3$C$_2$T$_x$ MXene nanosheet was also employed as the ETL in MAPbI$_3$-based perovskite solar cells [181]. In the work, the authors revealed that the UV-ozone treatment of Ti$_3$C$_2$T$_x$ induced oxide-like Ti–O bonds, resulting in enhanced electron transfer and suppressed nonradiative recombination at the Ti$_3$C$_2$T$_x$/MAPbI$_3$ interface. Attributed to this, the device showed improved performance with PCE significantly increased from 5% to 17.17%. The device performance of solar cells reviewed in Section 6.1 is summarized in Table 5.

Table 5. Summarized device performance for solar cells in Section 6.1.

Heterostructure	Voc (V)	Jsc (mA/cm^2)	FF (%)	PCE (%)	Ref.
GO/MAPbI$_{3-x}$Cl$_x$	1	18.4	76.8	14.14	[172]
rGO/MAPbI$_3$	0.962	22.1	77	16.4	[175]
rGO/MAPbI$_3$	1.08	18.06	77.7	15.2	[169]
GO/MAPbI$_3$	0.859	19.61	70.3	11.8	[176]
MoS$_2$/MAPbI$_3$	0.89	21.7	63.8	13.14	[170]
SnS$_2$/FA$_{0.75}$MA$_{0.15}$Cs$_{0.1}$PbI$_{2.65}$Br$_{0.35}$	1.161	23.55	73	20.12	[178]
MoS$_2$/MAPbI$_{3-x}$Cl$_x$	0.96	14.89	67	9.53	[179]
WS$_2$/MAPbI$_{3-x}$Cl$_x$	0.82	15.91	64	8.02	[179]
MoSe$_2$/MAPbI$_3$	1.02	14.45	55.8	8.23	[180]
BP/(FAPbI$_3$)$_x$(MAPbBr$_3$)$_{1-x}$	1.08	23.32	0.71	17.85	[177]
Ti$_3$C$_2$T$_x$ MXene/MAPbI$_3$	1.08	22.63	70	17.17	[181]

6.2. 2D material/Perovskite Heterostructure for Photodetectors

6.2.1. Graphene/Perovskite Heterostructure

Graphene has attracted much research interest in the field of photodetection owing to its high charge carrier mobility and broad absorption band. However, the development of graphene-based photodetectors has been largely restricted due to the rapid recombination rate and weak light absorption in pristine graphene. Various approaches have been applied to solve these problems, including coupling graphene with silicon waveguide [133] or microcavity [182], and hybridizing graphene with lead sulfide QDs [183]. However, some of these strategies require complicated fabrication procedures, or the obtained devices only have limited operating bandwidth and photosensitivity. To further improve the device performance, researchers integrated newly-emerged perovskite materials with graphene and great progress has been made in the past few years.

For 3D MAPbI$_3$ perovskite, the heterostructure with graphene demonstrated a high device performance [184]. Due to the formation of the heterostructure, electrons in the graphene layer are transported to the perovskite layer. These electrons filled the empty states caused by light absorption in the valence band, effectively reducing the probability of carrier recombination within the perovskite layer. Attributed to this photogating effect, the photodetector exhibited high photoresponsivity of 180 A/W, EQE of 5×10^4%, and detectivity $\approx 10^9$ Jones at light illumination power of 1 µW.

Recently, single-crystal MAPbBr$_3$ with the graphene heterostructure has also been demonstrated as shown in Figure 8a [185]. Enhanced charge transport and consequently high photocurrent were achieved in the device, which was attributed to low defect density in the single crystal perovskite and improved charge carrier separation by the internal electric field as shown in Figure 8b. Compared to devices based on pure graphene, the photoresponse of the graphene/MAPbI$_3$ heterostructure photodetector was enhanced by an order, with high responsivity (1017.1 A/W), detectivity (2.02×10^{13} Jones), and photoconductive gain (2.37×10^3).

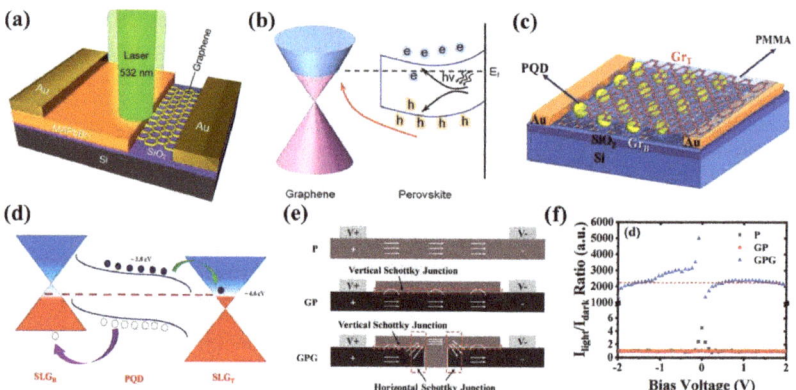

Figure 8. (a) Schematic of the graphene/MAPbI$_3$ single crystal photodetector. (b) Energy band diagram of the graphene/MAPbI$_3$ single crystal heterostructure. (c) Schematic of the graphene/PQDs/graphene phototransistor. (d) Energy band diagram of the graphene/PQDs/graphene heterostructure. (e) Schematic illustration of devices based on perovskite layer (P), graphene/perovskite (GP), and graphene/perovskite/graphene (GPG) structure. (f) The I$_{light}$/I$_{dark}$ ratio under different bias voltages. (a,b) Reprinted with permission from Ref. [185]. Copyright 2020, WILEY. (c,d) Reprinted with permission from Ref. [186]. Copyright 2019, ACS. (e,f) Reprinted with permission from Ref. [187]. Copyright 2019, WILEY.

In addition, graphene/perovskite/graphene (GPG) heterostructure was widely used for the photodetector application. For example, Bera et al. reported a broadband phototransistor based on the graphene/MAPbBr$_3$ PQDs/graphene heterostructure as shown in Figure 8c [186]. Due to the asymmetric potential at the interfaces (Figure 8d), electrons were transferred to the top graphene layer while holes were driven towards the bottom graphene layer under light illumination. The Schottky barrier suppressed carrier recombination and enhanced the photocurrent. In addition, the PMMA layer on the top of the device reduced the reflectivity of the incident light as well as served as an encapsulation layer against ambient air. The device exhibited high photoresponsivity of ~10^9 A/W, a fast response time of ~20 µs, and stable photodetection for over six months. Similarly, a photodetector consisting of the graphene/MAPbI$_3$/graphene structure was reported by Chen et al. in 2019 [187]. As shown in Figure 8e, in the GPG device, a perovskite channel was introduced in the bottom graphene layer, thus creating vertical and horizontal Schottky junctions for effective carrier transport. Compared to devices with other structures, the GPG device had the highest I$_{light}$/I$_{dark}$ ratio because the dark current was reduced by the perovskite channel with large resistance (Figure 8f).

The GPG phototransistor showed a high on/off switch ratio of 2.6×10^3 and detectivity of 3.55×10^9 Jones. The authors also fabricated the GPG structure on PET substrates and the obtained flexible device exhibited excellent reliability and bendability.

6.2.2. TMDC/Perovskite Heterostructure

A variety of TMDC/perovskite heterostructures have been employed for photodetector applications. For example, Lu et al. developed a high-performance phototransistor by integrating a WSe_2 monolayer and $MAPbI_3$ perovskite layer [188]. The authors used a focused laser beam to passivate the chalcogen vacancies in the pristine WSe_2 film and significantly improved its conductivity. The combination of perovskite with high absorption coefficient and defect-free WSe_2 resulted in excellent device performance which is three orders of magnitude higher than that of a pristine WSe_2 device.

Similarly, hybrid photodetectors comprising of $MoS_2/MAPbI_3$ heterostructure were proposed by employing different phases of MoS_2 layers, namely 1T and 2H [189]. The authors compared the performance of the metallic 1T phase and the semiconducting 2H phase of MoS_2 in photodetectors. On the one hand, the 1T phase is capable of accommodating the injection of more carriers owing to its metallic characteristic, resulting in enhanced photocurrent. On the other hand, the photogenerated carriers injected into MoS_2 have a high probability of recombination in the 1T-MoS_2/$MAPbI_3$ device (Figure 9a), which leads to a high dark current. As a result, the 1T-MoS_2/$MAPbI_3$ device exhibited high photoresponsivity with a low on/off ratio while the 2H-MoS_2/$MAPbI_3$ device had reasonable photoresponse with an increased on/off ratio.

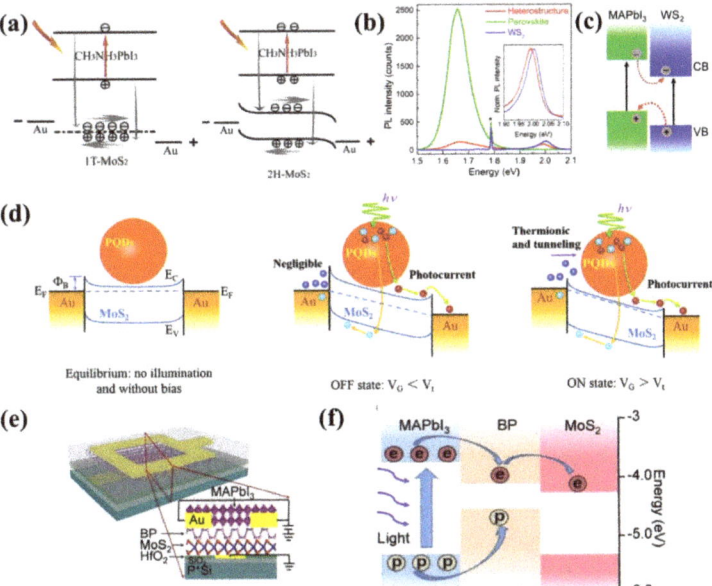

Figure 9. (a) Charge transfer mechanism of 1T-MoS_2/$MAPbI_3$ and 2H-MoS_2/$MAPbI_3$ photodetectors. (b) PL spectra of the fabricated heterostructure, isolated perovskite, and WS_2. (c) Schematic illustration of type II band alignment and charge transfer at the $MAPbI_3$/WS_2 heterostructure. (d) Schematic of charge transfer for the PQDs/MoS_2 device. Schematic illustration of the (e) perovskite/BP/MoS_2 photodiode and (f) charge transfer in the heterostructure. (a) Reprinted with permission from Ref. [189]. Copyright 2016, WILEY. (b,c) Reprinted with permission from Ref. [190]. Copyright 2019, ACS. (d) Reprinted with permission from Ref. [191]. Copyright 2018, WILEY. (e,f) Reprinted with permission from Ref. [192]. Copyright 2019, ACS.

In 2019, Erkōlōc et al. demonstrated photodetectors based on ultrathin MAPbI$_3$/WS$_2$ heterostructures by a vapor-phase growth approach [190]. PbI$_2$ was deposited onto monolayer WS$_2$, followed by the successive conversion to MAPbI$_3$ through MAI intercalation. WS$_2$ served as an effective template for the site-selective growth of ultrathin perovskite film at a large scale, avoiding any damage to the perovskite that would be caused by post-growth lithography processes. The strong PL quenching indicated effective charge transfer between MAPbI$_3$ and WS$_2$ layers due to type II band alignment (Figure 9b,c). Attributed to the strong light absorption capability of the perovskite layer, the fabricated photodetector showed enhanced responsivity of 43.6 A/W compared to the WS$_2$-only device of 3.3 A/W.

In addition to 3D perovskites, quasi-2D and PQDs have also been applied in heterostructures. For example, Fu et al. hybridized MoS$_2$ with quasi-2D (BA)$_2$(MA)$_2$Pb$_3$I$_{10}$ perovskite for sensitive photodetection [144]. Due to the facile charge separation at the MoS$_2$/perovskite interface, the hybrid photodetector exhibited excellent performance with two and six orders of magnitude improvement for detectivity (4×10^{10} Jones) and photoresponsivity (10^4 A/W). For PQDs, an ultrasensitive photodetector was demonstrated using the hybrid structure of CsPbI$_{3-x}$Br$_x$ QDs and monolayer MoS$_2$ [191]. Attributed to favorable energy band alignment at the PQDs/MoS$_2$ interface, the photocurrent of the hybrid device was improved by 15.3 times compared to that of the MoS$_2$-only device. As illustrated in Figure 9d, the Schottky barrier determined by the applied gate voltage and the photogating effect together modulated the dark current and photoresponsivity in the device. As a result, the optimized photodetector displayed high photoresponsivity of 7.7×10^4 A/W, detectivity of 5.6×10^{11} Jones and EQE of over 10^7%.

6.2.3. Other 2D Materials/Perovskite Heterostructure

Other 2D materials, such as BP and nitrides (MXenes), have also shown great potential for integrating with perovskites. For instance, Muduli et al. integrated CsPbBr$_3$ PQDs on the surface of few-layer BP (FLBP) [193]. Due to charge transfer between the CsPbBr$_3$ PQD and FLBP layers, the CsPbBr$_3$ PQD-FLBP device showed significant photoresponse in comparison with the CsPbBr$_3$ PQD-only device with negligible photocurrent. Following this work, a hybrid photodiode was proposed which consisted of the perovskite/BP/MoS$_2$ double heterostructures (Figure 9e) [192]. As illustrated in Figure 9f, type I and type II band alignments are formed at the perovskite/BP and BP/MoS$_2$ heterointerface, respectively. Under light illumination, photogenerated carriers in the perovskite layer diffused into the BP layer, followed by efficient carrier separation by the built-in electric field at the interface of BP and MoS$_2$. As a result, the photodiode exhibited high detectivity of 1.3×10^{12} Jones as well as a fast response of 150/240 µs.

For flexible photodetectors, a large-area photodetector was realized by utilizing a 2D CsPbBr$_3$ nanosheet as the light-absorbing material and MXene as electrodes, which was fabricated by all-sprayed-processable methods [194]. The obtained device showed a superior on/off ratio of 2.3×10^3 and a fast response speed of 18 ms, which was attributed to the highly conductive MXene electrodes, and excellent crystallinity of CsPbBr$_3$ nanosheet, and well-matched energy level at the CsPbBr$_3$/MXene interface. In addition, the device retained 85% of its original photocurrent after applying 1500 times of cyclic bending, demonstrating its excellent physical stability. The device performance of photodetectors reviewed in Section 6.2 is summarized in Table 6.

Table 6. Summarized device performance of photodetectors in Section 6.2.

Materials	Response Range	R (A/W)	D* (Jones)	τ_{rise}/τ_{fall}	Ref.
Graphene/MAPbI$_3$	UV–Visible	180	~10^9	87 ms/540 ms	[184]
Graphene/MAPbI$_3$	-	1017.1	2.02×10^{13}	50.9 ms/26 ms	[185]
Graphene/MAPbBr$_3$ QDs	UV–IR	~3×10^9	8.7×10^{15}	<50 µs/<50 µs	[186]
Graphene/MAPbI$_3$	650–900 nm	0.022	3.55×10^9	-	[187]
WSe$_2$/MAPbI$_3$	-	110	2.2×10^{11}	-	[188]
MoS$_2$/MAPbI$_3$	UV–Visible	3096	7×10^{11}	0.45 s/0.75 s	[189]
WS$_2$/MAPbI$_3$	-	43.6	-	-	[190]
MoS$_2$/(BA)$_2$(MA)$_2$Pb$_3$I$_{10}$	-	10^4	4×10^{10}	-	[144]
MoS$_2$/CsPbI$_{3-x}$Br$_x$	UV–Visible	7.7×10^4	5.6×10^{11}	0.59 s/0.32 s	[191]
BP/CsPbBr$_3$ QDs	-	-	1.3×10^{12}	150 µs/240 µs	[193]
MXene/CsPbBr$_3$	-	0.0449	6.4×10^8	48 ms/18 ms	[194]

7. Perspectives

To date, the 2D material/perovskite heterostructure has demonstrated great potential for advanced optoelectronics and numerous efforts have been made to improve the performance of devices. However, several challenges still remain which hinder the further improvement of commercial applications.

An open challenge is that the heterostructure generally suffers from poor stability in the ambient air because perovskite is very sensitive to humidity, heat, and light. In recent years, novel perovskites have been developed and demonstrated to be more stable than 3D perovskites. 2D RP perovskites have attracted much attention because their hydrophobic organic ligands could protect perovskites from water invasion [195]. In addition, it was reported that ion migration was suppressed in RP perovskites, which would alleviate the degradation process of perovskite [196]. Similarly, 2D DJ perovskites have emerged as a potential material with excellent environmental stability. Compared to RP perovskites, the adjacent inorganic layers of DJ perovskites were bridged by diammonium ligands which eliminate the vdWs gap and shorten the inter-slab distance, resulting in improved stability [197]. In addition, researchers have proposed a series of encapsulation materials for perovskite-based devices. For instance, Li et al. encapsulated CsPbBr$_3$ nanocrystals into dense SiO$_2$ solid and the PL intensity of perovskite maintained 100% of its initial value for 1000 h under illumination [198].

Apart from the stability issue, the large-scale production of 2D material/perovskite heterostructures has been limited by fabrication methods. On the one hand, the majority of high-quality heterostructures are constructed using solid-state methods, such as wet/dry transfer, mechanical exfoliation, CVD, and physical vapor deposition (PVD), which require complicated procedures and high cost. On the other hand, although facile solution-processed methods provide access to large yield, the heterostructures suffer from low quality and reduced charge transport due to the poor contact at the interface [33]. In this regard, epitaxial growth might be a promising strategy to solve this dilemma. It was demonstrated that this approach enabled precise control over the crystal phases, exposed interfaces, size, and morphology of the hybrid nanostructures with relatively low cost [199].

Lastly, the most widely researched perovskites in the heterostructures involve toxic and hazardous heavy metals, particularly Pb, which is harmful to the environment as well as human beings. Considerable efforts have been made toward replacing Pb with environmentally friendly or less hazardous elements, such as Sn, Mn, and Zn, where Sn is the most promising candidate among those substitutes [200–202]. However, Sn-based devices showed inferior performance in comparison with their lead counterparts because of the high density of defects in perovskite lattice, and severe oxidation of Sn from +2 to +4 states restricted their long-term stability. Therefore, the development of perovskites based on non-toxic substances is of high importance for future advanced optoelectronics.

8. Conclusions

The review covered the synthesis methods of 2D materials and perovskites, as well as their applications in solar cells and photodetectors. First, the basic properties and common synthesis routes of 2D materials and perovskites were addressed to provide a fundamental picture of these materials. Then, we summarized the recent research progress of optoelectronic devices based on 2D materials or perovskites. Despite their great potential in state-of-the-art optoelectronics, some issues regarding weak light absorption and poor stability restricted their commercial applications, which led to the introduction of 2D material/perovskite heterostructure. We analyzed the underlying benefits of this novel concept and reviewed the recent development of optoelectronic devices applying this heterostructure. Finally, we introduced the existing challenges and possible solutions for developing stable and high-quality 2D material/perovskite heterostructures, and also the need for environmentally friendly perovskites for green optoelectronics.

Author Contributions: S.M. and T.L. contributed equally to this work. T.L., Y.L., X.Z. and Y.X. did literature reviews for the materials section. S.M., Y.D., J.G. and F.S. did literature reviews for device parts. S.M. and T.L. wrote the manuscript under the supervision of Y.C. All authors have read and agreed to the published version of the manuscript.

Funding: This research received no external funding.

Data Availability Statement: Not applicable.

Acknowledgments: Y.C. thank UM-SJTU Joint Institute for its support of the work.

Conflicts of Interest: The authors declare no conflict of interest.

References

1. Ali, N.; Shehzad, N.; Uddin, S.; Ahmed, R.; Jabeen, M.; Kalam, A.; Al-Sehemi, A.G.; Alrobei, H.; Kanoun, M.B.; Khesro, A.; et al. A Review on Perovskite Materials with Solar Cell Prospective. *Int. J. Energy Res.* **2021**, *45*, 19729–19745. [CrossRef]
2. Li, L.; Ye, S.; Qu, J.; Zhou, F.; Song, J.; Shen, G. Recent Advances in Perovskite Photodetectors for Image Sensing. *Small* **2021**, *17*, 2005606. [CrossRef] [PubMed]
3. Liu, S.; Zhang, X.; Gu, X.; Ming, D. Photodetectors Based on Two Dimensional Materials for Biomedical Application. *Biosens. Bioelectron.* **2019**, *143*, 111617. [CrossRef] [PubMed]
4. Zhao, Z.; Liu, J.; Liu, Y.; Zhu, N. High-Speed Photodetectors in Optical Communication System. *J. Semicond.* **2017**, *38*, 121001. [CrossRef]
5. Mahapatra, A.; Prochowicz, D.; Tavakoli, M.M.; Trivedi, S.; Kumar, P.; Yadav, P. A Review of Aspects of Additive Engineering in Perovskite Solar Cells. *J. Mater. Chem. A* **2020**, *8*, 27–54. [CrossRef]
6. Ghosh, S.; Mishra, S.; Singh, T. Antisolvents in Perovskite Solar Cells: Importance, Issues, and Alternatives. *Adv. Mater. Interfaces* **2020**, *7*, 2000950. [CrossRef]
7. Li, X.; Hoffman, J.M.; Kanatzidis, M.G. The 2D Halide Perovskite Rulebook: How the Spacer Influences Everything from the Structure to Optoelectronic Device Efficiency. *Chem. Rev.* **2021**, *121*, 2230–2291. [CrossRef]
8. Zhang, F.; Lu, H.; Tong, J.; Berry, J.J.; Beard, M.C.; Zhu, K. Advances in Two-Dimensional Organic–Inorganic Hybrid Perovskites. *Energy Environ. Sci.* **2020**, *13*, 1154–1186. [CrossRef]
9. Chen, J.; Jia, D.; Johansson, E.M.J.; Hagfeldt, A.; Zhang, X. Emerging Perovskite Quantum Dot Solar Cells: Feasible Approaches to Boost Performance. *Energy Environ. Sci.* **2021**, *14*, 224–261. [CrossRef]
10. Kolobov, A.V.; Tominaga, J. *Two-Dimensional Transition-Metal Dichalcogenides*; Springer Series in Materials Science; Springer International Publishing: Cham, Switzerland, 2016; Volume 239, ISBN 978-3-319-31449-5.
11. Shi, Y.; Li, H.; Li, L.-J. Recent Advances in Controlled Synthesis of Two-Dimensional Transition Metal Dichalcogenides via Vapour Deposition Techniques. *Chem. Soc. Rev.* **2015**, *44*, 2744–2756. [CrossRef]
12. Whitener, K.E.; Sheehan, P.E. Graphene Synthesis. *Diam. Relat. Mater.* **2014**, *46*, 25–34. [CrossRef]
13. Chen, S.; Shi, G. Two-Dimensional Materials for Halide Perovskite-Based Optoelectronic Devices. *Adv. Mater.* **2017**, *29*, 1605448. [CrossRef] [PubMed]
14. Raja, S.N.; Bekenstein, Y.; Koc, M.A.; Fischer, S.; Zhang, D.; Lin, L.; Ritchie, R.O.; Yang, P.; Alivisatos, A.P. Encapsulation of Perovskite Nanocrystals into Macroscale Polymer Matrices: Enhanced Stability and Polarization. *ACS Appl. Mater. Interfaces* **2016**, *8*, 35523–35533. [CrossRef] [PubMed]
15. Konidakis, I.; Karagiannaki, A.; Stratakis, E. Advanced Composite Glasses with Metallic, Perovskite, and Two-Dimensional Nanocrystals for Optoelectronic and Photonic Applications. *Nanoscale* **2022**, *14*, 2966–2989. [CrossRef] [PubMed]

16. Lee, C.; Wei, X.; Kysar, J.; Hone, J. Measurement of the Elastic Properties and Intrinsic Strength of Monolayer Graphene. *Science* **2008**, *321*, 382–385. [CrossRef] [PubMed]
17. Balandin, A.A.; Ghosh, S.; Bao, W.; Calizo, I.; Teweldebrhan, D.; Miao, F.; Lau, C.N. Superior Thermal Conductivity of Single-Layer Graphene. *Nano Lett.* **2008**, *8*, 902–907. [CrossRef]
18. Xu, Y.; Shi, Z.; Shi, X.; Zhang, K.; Zhang, H. Recent Progress in Black Phosphorus and Black-Phosphorus-Analogue Materials: Properties, Synthesis and Applications. *Nanoscale* **2019**, *11*, 14491–14527. [CrossRef]
19. Ma, X.; Wu, X.; Wang, H.; Wang, Y. A Janus MoSSe Monolayer: A Potential Wide Solar-Spectrum Water-Splitting Photocatalyst with a Low Carrier Recombination Rate. *J. Mater. Chem. A* **2018**, *6*, 2295–2301. [CrossRef]
20. Eda, G.; Yamaguchi, H.; Voiry, D.; Fujita, T.; Chen, M.; Chhowalla, M. Photoluminescence from Chemically Exfoliated MoS_2. *Nano Lett.* **2011**, *11*, 5111–5116. [CrossRef]
21. Novoselov, K.S.; Geim, A.K.; Morozov, S.V.; Jiang, D.; Zhang, Y.; Dubonos, S.V.; Grigorieva, I.V.; Firsov, A.A. Electric Field Effect in Atomically Thin Carbon Films. *Science* **2004**, *306*, 666–669. [CrossRef]
22. Magda, G.Z.; Pető, J.; Dobrik, G.; Hwang, C.; Biró, L.P.; Tapasztó, L. Exfoliation of Large-Area Transition Metal Chalcogenide Single Layers. *Sci. Rep.* **2015**, *5*, 14714. [CrossRef] [PubMed]
23. Dines, M.B. Lithium Intercalation via N-Butyl Lithium of the Layered Transition Metal Dichalcogenides. *Mater. Res. Bull.* **1975**, *10*, 5. [CrossRef]
24. Li, L.; Yu, Y.; Ye, G.J.; Ge, Q.; Ou, X.; Wu, H.; Feng, D.; Chen, X.H.; Zhang, Y. Black Phosphorus Field-Effect Transistors. *Nat. Nanotechnol.* **2014**, *9*, 372–377. [CrossRef] [PubMed]
25. Eda, G.; Maier, S.A. Two-Dimensional Crystals: Managing Light for Optoelectronics. *ACS Nano* **2013**, *7*, 5660–5665. [CrossRef]
26. Kojima, A.; Teshima, K.; Shirai, Y.; Miyasaka, T. Organometal Halide Perovskites as Visible-Light Sensitizers for Photovoltaic Cells. *J. Am. Chem. Soc.* **2009**, *131*, 6050–6051. [CrossRef]
27. Jeong, J.; Kim, M.; Seo, J.; Lu, H.; Ahlawat, P.; Mishra, A.; Yang, Y.; Hope, M.A.; Eickemeyer, F.T.; Kim, M.; et al. Pseudo-Halide Anion Engineering for α-$FAPbI_3$ Perovskite Solar Cells. *Nature* **2021**, *592*, 381–385. [CrossRef]
28. Yang, Z.; Wang, M.; Li, J.; Dou, J.; Qiu, H.; Shao, J. Spray-Coated $CsPbBr_3$ Quantum Dot Films for Perovskite Photodiodes. *ACS Appl. Mater. Interfaces* **2018**, *10*, 26387–26395. [CrossRef]
29. Liang, F.-X.; Wang, J.-Z.; Zhang, Z.-X.; Wang, Y.-Y.; Gao, Y.; Luo, L.-B. Broadband, Ultrafast, Self-Driven Photodetector Based on Cs-Doped $FAPbI_3$ Perovskite Thin Film. *Adv. Opt. Mater.* **2017**, *5*, 1700654. [CrossRef]
30. D'Innocenzo, V.; Grancini, G.; Alcocer, M.J.P.; Kandada, A.R.S.; Stranks, S.D.; Lee, M.M.; Lanzani, G.; Snaith, H.J.; Petrozza, A. Excitons versus Free Charges in Organo-Lead Tri-Halide Perovskites. *Nat. Commun.* **2014**, *5*, 3586. [CrossRef]
31. Stranks, S.D.; Eperon, G.E.; Grancini, G.; Menelaou, C.; Alcocer, M.J.P.; Leijtens, T.; Herz, L.M.; Petrozza, A.; Snaith, H.J. Electron-Hole Diffusion Lengths Exceeding 1 Micrometer in an Organometal Trihalide Perovskite Absorber. *Science* **2013**, *342*, 341–344. [CrossRef]
32. Yang, X.; Luo, D.; Xiang, Y.; Zhao, L.; Anaya, M.; Shen, Y.; Wu, J.; Yang, W.; Chiang, Y.-H.; Tu, Y.; et al. Buried Interfaces in Halide Perovskite Photovoltaics. *Adv. Mater* **2021**, *10*, 2006435. [CrossRef] [PubMed]
33. Zhang, Z.; Wang, S.; Liu, X.; Chen, Y.; Su, C.; Tang, Z.; Li, Y.; Xing, G. Metal Halide Perovskite/2D Material Heterostructures: Syntheses and Applications. *Small Methods* **2021**, *5*, 2000937. [CrossRef] [PubMed]
34. Bolotin, K.I.; Sikes, K.J.; Jiang, Z.; Klima, M.; Fudenberg, G.; Hone, J.; Kim, P.; Stormer, H.L. Ultrahigh Electron Mobility in Suspended Graphene. *Solid State Commun.* **2008**, *146*, 351–355. [CrossRef]
35. Novoselov, K.S.; Geim, A.K.; Morozov, S.V.; Jiang, D.; Katsnelson, M.I.; Grigorieva, I.V.; Dubonos, S.V.; Firsov, A.A. Two-Dimensional Gas of Massless Dirac Fermions in Graphene. *Nature* **2005**, *438*, 197–200. [CrossRef] [PubMed]
36. Zhang, Y.; Tan, Y.-W.; Stormer, H.L.; Kim, P. Experimental Observation of the Quantum Hall Effect and Berry's Phase in Graphene. *Nature* **2005**, *438*, 201–204. [CrossRef] [PubMed]
37. Semenoff, G.W. Condensed-Matter Simulation of a Three-Dimensional Anomaly. *Phys. Rev. Lett.* **1984**, *53*, 2449–2452. [CrossRef]
38. Novoselov, K.S. Nobel Lecture: Graphene: Materials in the Flatland. *Rev. Mod. Phys.* **2011**, *83*, 837–849. [CrossRef]
39. Chua, C.K.; Pumera, M. Chemical Reduction of Graphene Oxide: A Synthetic Chemistry Viewpoint. *Chem. Soc. Rev.* **2014**, *43*, 291–312. [CrossRef]
40. Chandrashekar, B.N.; Deng, B.; Smitha, A.S.; Chen, Y.; Tan, C.; Zhang, H.; Peng, H.; Liu, Z. Roll-to-Roll Green Transfer of CVD Graphene onto Plastic for a Transparent and Flexible Triboelectric Nanogenerator. *Adv. Mater.* **2015**, *27*, 5210–5216. [CrossRef]
41. Moon, I.K.; Lee, J.; Ruoff, R.S.; Lee, H. Reduced Graphene Oxide by Chemical Graphitization. *Nat. Commun.* **2010**, *1*, 73. [CrossRef]
42. Ambrosi, A.; Chua, C.K.; Bonanni, A.; Pumera, M. Lithium Aluminum Hydride as Reducing Agent for Chemically Reduced Graphene Oxides. *Chem. Mater.* **2012**, *24*, 2292–2298. [CrossRef]
43. Feng, H.; Cheng, R.; Zhao, X.; Duan, X.; Li, J. A Low-Temperature Method to Produce Highly Reduced Graphene Oxide. *Nat. Commun.* **2013**, *4*, 1539. [CrossRef] [PubMed]
44. Li, X.; Cai, W.; An, J.; Kim, S.; Nah, J.; Yang, D.; Piner, R.; Velamakanni, A.; Jung, I.; Tutuc, E.; et al. Large-Area Synthesis of High-Quality and Uniform Graphene Films on Copper Foils. *Science* **2009**, *324*, 1312–1314. [CrossRef] [PubMed]
45. Kim, K.S.; Zhao, Y.; Jang, H.; Lee, S.Y.; Kim, J.M.; Kim, K.S.; Ahn, J.-H.; Kim, P.; Choi, J.-Y.; Hong, B.H. Large-Scale Pattern Growth of Graphene Films for Stretchable Transparent Electrodes. *Nature* **2009**, *457*, 706–710. [CrossRef]

46. Bae, S.; Kim, H.; Lee, Y.; Xu, X.; Park, J.-S.; Zheng, Y.; Balakrishnan, J.; Lei, T.; Ri Kim, H.; Song, Y.I.; et al. Roll-to-Roll Production of 30-Inch Graphene Films for Transparent Electrodes. *Nat. Nanotechnol.* **2010**, *5*, 574–578. [CrossRef]
47. Bosi, M. Growth and Synthesis of Mono and Few-Layers Transition Metal Dichalcogenides by Vapour Techniques: A Review. *RSC Adv.* **2015**, *5*, 75500–75518. [CrossRef]
48. Yue, R.; Barton, A.T.; Zhu, H.; Azcatl, A.; Pena, L.F.; Wang, J.; Peng, X.; Lu, N.; Cheng, L.; Addou, R.; et al. HfSe$_2$ Thin Films: 2D Transition Metal Dichalcogenides Grown by Molecular Beam Epitaxy. *ACS Nano* **2015**, *9*, 474–480. [CrossRef]
49. Zeng, Z.; Yin, Z.; Huang, X.; Li, H.; He, Q.; Lu, G.; Boey, F.; Zhang, H. Single-Layer Semiconducting Nanosheets: High-Yield Preparation and Device Fabrication. *Angew. Chem.* **2011**, *123*, 11289–11293. [CrossRef]
50. Li, H.; Wu, J.; Yin, Z.; Zhang, H. Preparation and Applications of Mechanically Exfoliated Single-Layer and Multilayer MoS$_2$ and WSe$_2$ Nanosheets. *Acc. Chem. Res.* **2014**, *47*, 1067–1075. [CrossRef]
51. Jeong, S.; Yoo, D.; Ahn, M.; Miró, P.; Heine, T.; Cheon, J. Tandem Intercalation Strategy for Single-Layer Nanosheets as an Effective Alternative to Conventional Exfoliation Processes. *Nat. Commun.* **2015**, *6*, 5763. [CrossRef]
52. Raza, A.; Hassan, J.Z.; Ikram, M.; Ali, S.; Farooq, U.; Khan, Q.; Maqbool, M. Advances in Liquid-Phase and Intercalation Exfoliations of Transition Metal Dichalcogenides to Produce 2D Framework. *Adv. Mater. Interfaces* **2021**, *8*, 2002205. [CrossRef]
53. Coleman, J.N.; Lotya, M.; O'Neill, A.; Bergin, S.D.; King, P.J.; Khan, U.; Young, K.; Gaucher, A.; De, S.; Smith, R.J.; et al. Two-Dimensional Nanosheets Produced by Liquid Exfoliation of Layered Materials. *Science* **2011**, *331*, 568–571. [CrossRef] [PubMed]
54. Chang, Y.-H.; Zhang, W.; Zhu, Y.; Han, Y.; Pu, J.; Chang, J.-K.; Hsu, W.-T.; Huang, J.-K.; Hsu, C.-L.; Chiu, M.-H.; et al. Monolayer MoSe$_2$ Grown by Chemical Vapor Deposition for Fast Photodetection. *ACS Nano* **2014**, *8*, 8582–8590. [CrossRef] [PubMed]
55. Zhang, Y.; Zhang, Y.; Ji, Q.; Ju, J.; Yuan, H.; Shi, J.; Gao, T.; Ma, D.; Liu, M.; Chen, Y.; et al. Controlled Growth of High-Quality Monolayer WS$_2$ Layers on Sapphire and Imaging Its Grain Boundary. *ACS Nano* **2013**, *7*, 8963–8971. [CrossRef]
56. Liu, B.; Fathi, M.; Chen, L.; Abbas, A.; Ma, Y.; Zhou, C. Chemical Vapor Deposition Growth of Monolayer WSe$_2$ with Tunable Device Characteristics and Growth Mechanism Study. *ACS Nano* **2015**, *9*, 6119–6127. [CrossRef]
57. Orofeo, C.M.; Suzuki, S.; Sekine, Y.; Hibino, H. Scalable Synthesis of Layer-Controlled WS$_2$ and MoS$_2$ Sheets by Sulfurization of Thin Metal Films. *Appl. Phys. Lett.* **2014**, *105*, 083112. [CrossRef]
58. Jin, Z.; Shin, S.; Kwon, D.H.; Han, S.-J.; Min, Y.-S. Novel Chemical Route for Atomic Layer Deposition of MoS$_2$ Thin Film on SiO$_2$/Si Substrate. *Nanoscale* **2014**, *6*, 14453–14458. [CrossRef]
59. Tan, Z.-K.; Moghaddam, R.S.; Lai, M.L.; Docampo, P.; Higler, R.; Deschler, F.; Price, M.; Sadhanala, A.; Pazos, L.M.; Credgington, D.; et al. Bright Light-Emitting Diodes Based on Organometal Halide Perovskite. *Nat. Nanotechnol.* **2014**, *9*, 687–692. [CrossRef]
60. Fang, Y.; Dong, Q.; Shao, Y.; Yuan, Y.; Huang, J. Highly Narrowband Perovskite Single-Crystal Photodetectors Enabled by Surface-Charge Recombination. *Nat. Photonics* **2015**, *9*, 679–686. [CrossRef]
61. Zhao, W.; Lin, H.; Li, Y.; Wang, D.; Wang, J.; Liu, Z.; Yuan, N.; Ding, J.; Wang, Q.; Liu, S. Symmetrical Acceptor–Donor–Acceptor Molecule as a Versatile Defect Passivation Agent toward Efficient FA$_{0.85}$MA$_{0.15}$PbI$_3$ Perovskite Solar Cells. *Adv. Funct. Mater.* **2022**, *32*, 2112032. [CrossRef]
62. Han, Y.; Xie, H.; Lim, E.L.; Bi, D. Review of Two-Step Method for Lead Halide Perovskite Solar Cells. *Sol. RRL* **2022**, *6*, 2101007. [CrossRef]
63. Vaynzof, Y. The Future of Perovskite Photovoltaics—Thermal Evaporation or Solution Processing? *Adv. Energy Mater.* **2020**, *10*, 2003073. [CrossRef]
64. Chiang, C.-H.; Lin, J.-W.; Wu, C.-G. One-Step Fabrication of a Mixed-Halide Perovskite Film for a High-Efficiency Inverted Solar Cell and Module. *J. Mater. Chem. A* **2016**, *4*, 13525–13533. [CrossRef]
65. Chen, Q.; Deng, K.; Shen, Y.; Li, L. Stable One Dimensional (1D)/Three Dimensional (3D) Perovskite Solar Cell with an Efficiency Exceeding 23%. *InfoMat* **2022**, *4*, e12303. [CrossRef]
66. Wang, J.; Lin, D.; Chen, Y.; Luo, S.; Ke, L.; Ren, X.; Cui, S.; Zhang, L.; Li, Z.; Meng, K.; et al. Suppressing the Excessive Solvated Phase for Dion–Jacobson Perovskites with Improved Crystallinity and Vertical Orientation. *Sol. RRL* **2020**, *4*, 2000371. [CrossRef]
67. Zhang, F.; Zhong, H.; Chen, C.; Wu, X.; Hu, X.; Huang, H.; Han, J.; Zou, B.; Dong, Y. Brightly Luminescent and Color-Tunable Colloidal CH$_3$NH$_3$PbX$_3$ (X = Br, I, Cl) Quantum Dots: Potential Alternatives for Display Technology. *ACS Nano* **2015**, *9*, 4533–4542. [CrossRef]
68. Protesescu, L.; Yakunin, S.; Bodnarchuk, M.I.; Krieg, F.; Caputo, R.; Hendon, C.H.; Yang, R.X.; Walsh, A.; Kovalenko, M.V. Nanocrystals of Cesium Lead Halide Perovskites (CsPbX$_3$, X = Cl, Br, and I): Novel Optoelectronic Materials Showing Bright Emission with Wide Color Gamut. *Nano Lett.* **2015**, *15*, 3692–3696. [CrossRef]
69. De Roo, J.; Ibáñez, M.; Geiregat, P.; Nedelcu, G.; Walravens, W.; Maes, J.; Martins, J.C.; Van Driessche, I.; Kovalenko, M.V.; Hens, Z. Highly Dynamic Ligand Binding and Light Absorption Coefficient of Cesium Lead Bromide Perovskite Nanocrystals. *ACS Nano* **2016**, *10*, 2071–2081. [CrossRef]
70. Soe, C.M.M.; Stoumpos, C.C.; Kepenekian, M.; Traoré, B.; Tsai, H.; Nie, W.; Wang, B.; Katan, C.; Seshadri, R.; Mohite, A.D.; et al. New Type of 2D Perovskites with Alternating Cations in the Interlayer Space, (C(NH$_2$)$_3$)(CH$_3$NH$_3$)$_n$Pb$_n$I$_{3n+1}$: Structure, Properties, and Photovoltaic Performance. *J. Am. Chem. Soc.* **2017**, *139*, 16297–16309. [CrossRef]
71. Gong, X.; Guan, L.; Pan, H.; Sun, Q.; Zhao, X.; Li, H.; Pan, H.; Shen, Y.; Shao, Y.; Sun, L.; et al. Highly Efficient Perovskite Solar Cells via Nickel Passivation. *Adv. Funct. Mater.* **2018**, *28*, 1804286. [CrossRef]

72. Zhi, L.; Li, Y.; Cao, X.; Li, Y.; Cui, X.; Ci, L.; Wei, J. Dissolution and Recrystallization of Perovskite Induced by N-Methyl-2-Pyrrolidone in a Closed Steam Annealing Method. *J. Energy Chem.* **2019**, *30*, 78–83. [CrossRef]
73. Noel, N.K.; Abate, A.; Stranks, S.D.; Parrott, E.S.; Burlakov, V.M.; Goriely, A.; Snaith, H.J. Enhanced Photoluminescence and Solar Cell Performance via Lewis Base Passivation of Organic–Inorganic Lead Halide Perovskites. *ACS Nano* **2014**, *8*, 9815–9821. [CrossRef] [PubMed]
74. Pham, H.T.; Yin, Y.; Andersson, G.; Weber, K.J.; Duong, T.; Wong-Leung, J. Unraveling the Influence of CsCl/MACl on the Formation of Nanotwins, Stacking Faults and Cubic Supercell Structure in FA-Based Perovskite Solar Cells. *Nano Energy* **2021**, *87*, 106226. [CrossRef]
75. Yang, L.; Gao, Y.; Wu, Y.; Xue, X.; Wang, F.; Sui, Y.; Sun, Y.; Wei, M.; Liu, X.; Liu, H. Novel Insight into the Role of Chlorobenzene Antisolvent Engineering for Highly Efficient Perovskite Solar Cells: Gradient Diluted Chlorine Doping. *ACS Appl. Mater. Interfaces* **2019**, *11*, 792–801. [CrossRef] [PubMed]
76. Xiao, M.; Zhao, L.; Geng, M.; Li, Y.; Dong, B.; Xu, Z.; Wan, L.; Li, W.; Wang, S. Selection of an Anti-Solvent for Efficient and Stable Cesium-Containing Triple Cation Planar Perovskite Solar Cells. *Nanoscale* **2018**, *10*, 12141–12148. [CrossRef]
77. Jeon, N.J.; Noh, J.H.; Kim, Y.C.; Yang, W.S.; Ryu, S.; Seok, S.I. Solvent Engineering for High-Performance Inorganic–Organic Hybrid Perovskite Solar Cells. *Nat. Mater.* **2014**, *13*, 897–903. [CrossRef] [PubMed]
78. Zhang, W. Ethyl Acetate Green Antisolvent Process for High-Performance Planar Low-Temperature SnO$_2$-Based Perovskite Solar Cells Made in Ambient Air. *Chem. Eng. J.* **2020**, *79*, 122298. [CrossRef]
79. Taylor, A.D.; Sun, Q.; Goetz, K.P.; An, Q.; Schramm, T.; Hofstetter, Y.; Litterst, F.; Paulus, F.; Vaynzof, Y. A General Approach to High-Efficiency Perovskite Solar Cells by Any Antisolvent. *Nat. Commun.* **2021**, *12*, 1878. [CrossRef]
80. Lohmann, K.B.; Patel, J.B.; Rothmann, M.U.; Xia, C.Q.; Oliver, R.D.J.; Herz, L.M.; Snaith, H.J.; Johnston, M.B. Control over Crystal Size in Vapor Deposited Metal-Halide Perovskite Films. *ACS Energy Lett.* **2020**, *5*, 710–717. [CrossRef]
81. Juarez-Perez, E.J.; Hawash, Z.; Raga, S.R.; Ono, L.K.; Qi, Y. Thermal Degradation of CH$_3$NH$_3$PbI$_3$ Perovskite into NH$_3$ and CH$_3$I Gases Observed by Coupled Thermogravimetry–Mass Spectrometry Analysis. *Energy Environ. Sci.* **2016**, *9*, 3406–3410. [CrossRef]
82. Lohmann, K.B.; Motti, S.G.; Oliver, R.D.J.; Ramadan, A.J.; Sansom, H.C.; Yuan, Q.; Elmestekawy, K.A.; Patel, J.B.; Ball, J.M.; Herz, L.M.; et al. Solvent-Free Method for Defect Reduction and Improved Performance of p-i-n Vapor-Deposited Perovskite Solar Cells. *ACS Energy Lett.* **2022**, *7*, 1903–1911. [CrossRef]
83. Arciniegas, M.P.; Castelli, A.; Piazza, S.; Dogan, S.; Ceseracciu, L.; Krahne, R.; Duocastella, M.; Manna, L. Laser-Induced Localized Growth of Methylammonium Lead Halide Perovskite Nano- and Microcrystals on Substrates. *Adv. Funct. Mater.* **2017**, *27*, 1701613. [CrossRef]
84. Zou, C.; Zheng, J.; Chang, C.; Majumdar, A.; Lin, L.Y. Nonvolatile Rewritable Photomemory Arrays Based on Reversible Phase-Change Perovskite for Optical Information Storage. *Adv. Opt. Mater.* **2019**, *7*, 1900558. [CrossRef]
85. Mao, L.; Ke, W.; Pedesseau, L.; Wu, Y.; Katan, C.; Even, J.; Wasielewski, M.R.; Stoumpos, C.C.; Kanatzidis, M.G. Hybrid Dion–Jacobson 2D Lead Iodide Perovskites. *J. Am. Chem. Soc.* **2018**, *140*, 3775–3783. [CrossRef] [PubMed]
86. Billing, D.G.; Lemmerer, A. Synthesis, Characterization and Phase Transitions in the Inorganic–Organic Layered Perovskite-Type Hybrids [(C$_n$H$_{2n+1}$NH$_3$)$_2$PbI$_4$], n = 4, 5 and 6. *Acta Crystallogr. B* **2007**, *63*, 735–747. [CrossRef] [PubMed]
87. Li, X.; Hoffman, J.; Ke, W.; Chen, M.; Tsai, H.; Nie, W.; Mohite, A.D.; Kepenekian, M.; Katan, C.; Even, J.; et al. Two-Dimensional Halide Perovskites Incorporating Straight Chain Symmetric Diammonium Ions, (NH$_3$C$_m$H$_{2m}$NH$_3$)(CH$_3$NH$_3$)$_{n-1}$Pb$_n$I$_{3n+1}$ (m = 4–9; n = 1–4). *J. Am. Chem. Soc.* **2018**, *140*, 12226–12238. [CrossRef]
88. Lemmerer, A.; Billing, D.G. Lead Halide Inorganic–Organic Hybrids Incorporating Diammonium Cations. *CrystEngComm* **2012**, *14*, 1954. [CrossRef]
89. Mao, L.; Stoumpos, C.C.; Kanatzidis, M.G. Two-Dimensional Hybrid Halide Perovskites: Principles and Promises. *J. Am. Chem. Soc.* **2019**, *141*, 1171–1190. [CrossRef]
90. Lu, D.; Lv, G.; Xu, Z.; Dong, Y.; Ji, X.; Liu, Y. Thiophene-Based Two-Dimensional Dion–Jacobson Perovskite Solar Cells with over 15% Efficiency. *J. Am. Chem. Soc.* **2020**, *142*, 11114–11122. [CrossRef]
91. Stoumpos, C.C.; Soe, C.M.M.; Tsai, H.; Nie, W.; Blancon, J.-C.; Cao, D.H.; Liu, F.; Traoré, B.; Katan, C.; Even, J.; et al. High Members of the 2D Ruddlesden-Popper Halide Perovskites: Synthesis, Optical Properties, and Solar Cells of (CH$_3$(CH$_2$)$_3$NH$_3$)$_2$(CH$_3$NH$_3$)$_4$Pb$_5$I$_{16}$. *Chem* **2017**, *2*, 427–440. [CrossRef]
92. Zhang, X.; Wu, G.; Fu, W.; Qin, M.; Yang, W.; Yan, J.; Zhang, Z.; Lu, X.; Chen, H. Orientation Regulation of Phenylethylammonium Cation Based 2D Perovskite Solar Cell with Efficiency Higher Than 11%. *Adv. Energy Mater.* **2018**, *8*, 1702498. [CrossRef]
93. Marimuthu, T.; Yuvakkumar, R.; Kumar, P.S.; Vo, D.-V.N.; Xu, X.; Xu, G. Two-Dimensional Hybrid Perovskite Solar Cells: A Review. *Environ. Chem. Lett.* **2022**, *20*, 189–210. [CrossRef]
94. Liang, D.; Dong, C.; Cai, L.; Su, Z.; Zang, J.; Wang, C.; Wang, X.; Zou, Y.; Li, Y.; Chen, L.; et al. Unveiling Crystal Orientation in Quasi-2D Perovskite Films by In Situ GIWAXS for High-Performance Photovoltaics. *Small* **2021**, *17*, 2100972. [CrossRef] [PubMed]
95. Quan, L.N.; Zhao, Y.; García de Arquer, F.P.; Sabatini, R.; Walters, G.; Voznyy, O.; Comin, R.; Li, Y.; Fan, J.Z.; Tan, H.; et al. Tailoring the Energy Landscape in Quasi-2D Halide Perovskites Enables Efficient Green-Light Emission. *Nano Lett.* **2017**, *17*, 3701–3709. [CrossRef]
96. Ma, D.; Lin, K.; Dong, Y.; Choubisa, H.; Proppe, A.H.; Wu, D.; Wang, Y.-K.; Chen, B.; Li, P.; Fan, J.Z.; et al. Distribution Control Enables Efficient Reduced-Dimensional Perovskite LEDs. *Nature* **2021**, *599*, 594–598. [CrossRef]

97. Proppe, A.H.; Wei, M.; Chen, B.; Quintero-Bermudez, R.; Kelley, S.O.; Sargent, E.H. Photochemically Cross-Linked Quantum Well Ligands for 2D/3D Perovskite Photovoltaics with Improved Photovoltage and Stability. *J. Am. Chem. Soc.* **2019**, *141*, 14180–14189. [CrossRef]
98. Teale, S.; Proppe, A.H.; Jung, E.H.; Johnston, A.; Parmar, D.H.; Chen, B.; Hou, Y.; Kelley, S.O.; Sargent, E.H. Dimensional Mixing Increases the Efficiency of 2D/3D Perovskite Solar Cells. *J. Phys. Chem. Lett.* **2020**, *11*, 5115–5119. [CrossRef]
99. Chen, H.; Teale, S.; Chen, B.; Hou, Y.; Grater, L.; Zhu, T.; Bertens, K.; Park, S.M.; Atapattu, H.R.; Gao, Y.; et al. Quantum-Size-Tuned Heterostructures Enable Efficient and Stable Inverted Perovskite Solar Cells. *Nat. Photonics* **2022**, *16*, 352–358. [CrossRef]
100. Liu, C.; Sun, J.; Tan, W.L.; Lu, J.; Gengenbach, T.R.; McNeill, C.R.; Ge, Z.; Cheng, Y.-B.; Bach, U. Alkali Cation Doping for Improving the Structural Stability of 2D Perovskite in 3D/2D PSCs. *Nano Lett.* **2020**, *20*, 1240–1251. [CrossRef]
101. Yoo, J.J.; Wieghold, S.; Sponseller, M.C.; Chua, M.R.; Bertram, S.N.; Hartono, N.T.P.; Tresback, J.S.; Hansen, E.C.; Correa-Baena, J.-P.; Bulović, V.; et al. An Interface Stabilized Perovskite Solar Cell with High Stabilized Efficiency and Low Voltage Loss. *Energy Environ. Sci.* **2019**, *12*, 2192–2199. [CrossRef]
102. Jang, Y.-W.; Lee, S.; Yeom, K.M.; Jeong, K.; Choi, K.; Choi, M.; Noh, J.H. Intact 2D/3D Halide Junction Perovskite Solar Cells via Solid-Phase in-Plane Growth. *Nat. Energy* **2021**, *6*, 63–71. [CrossRef]
103. Huang, X.; Guo, Q.; Yang, D.; Xiao, X.; Liu, X.; Xia, Z.; Fan, F.; Qiu, J.; Dong, G. Reversible 3D Laser Printing of Perovskite Quantum Dots inside a Transparent Medium. *Nat. Photonics* **2020**, *14*, 82–88. [CrossRef]
104. Schmidt, L.C.; Pertegás, A.; González-Carrero, S.; Malinkiewicz, O.; Agouram, S.; Mínguez Espallargas, G.; Bolink, H.J.; Galian, R.E.; Pérez-Prieto, J. Nontemplate Synthesis of $CH_3NH_3PbBr_3$ Perovskite Nanoparticles. *J. Am. Chem. Soc.* **2014**, *136*, 850–853. [CrossRef] [PubMed]
105. Gonzalez-Carrero, S.; Galian, R.E.; Pérez-Prieto, J. Maximizing the Emissive Properties of $CH_3NH_3PbBr_3$ Perovskite Nanoparticles. *J. Mater. Chem. A* **2015**, *3*, 9187–9193. [CrossRef]
106. Yettapu, G.R.; Talukdar, D.; Sarkar, S.; Swarnkar, A.; Nag, A.; Ghosh, P.; Mandal, P. Terahertz Conductivity within Colloidal $CsPbBr_3$ Perovskite Nanocrystals: Remarkably High Carrier Mobilities and Large Diffusion Lengths. *Nano Lett.* **2016**, *16*, 4838–4848. [CrossRef]
107. Zou, T.; Liu, X.; Qiu, R.; Wang, Y.; Huang, S.; Liu, C.; Dai, Q.; Zhou, H. Enhanced UV-C Detection of Perovskite Photodetector Arrays via Inorganic $CsPbBr_3$ Quantum Dot Down-Conversion Layer. *Adv. Opt. Mater.* **2019**, *7*, 1801812. [CrossRef]
108. Lu, J.; Sheng, X.; Tong, G.; Yu, Z.; Sun, X.; Yu, L.; Xu, J.; Wang, J.; Xu, J.; Shi, Y.; et al. Ultrafast Solar-Blind Ultraviolet Detection by Inorganic Perovskite $CsPbX_3$ Quantum Dots Radial Junction Architecture. *Adv. Mater.* **2017**, *29*, 1700400. [CrossRef]
109. Li, J.; Xu, L.; Wang, T.; Song, J.; Chen, J.; Xue, J.; Dong, Y.; Cai, B.; Shan, Q.; Han, B.; et al. 50-Fold EQE Improvement up to 6.27% of Solution-Processed All-Inorganic Perovskite $CsPbBr_3$ QLEDs via Surface Ligand Density Control. *Adv. Mater.* **2017**, *29*, 1603885. [CrossRef]
110. Yan, F.; Xing, J.; Xing, G.; Quan, L.; Tan, S.T.; Zhao, J.; Su, R.; Zhang, L.; Chen, S.; Zhao, Y.; et al. Highly Efficient Visible Colloidal Lead-Halide Perovskite Nanocrystal Light-Emitting Diodes. *Nano Lett.* **2018**, *18*, 3157–3164. [CrossRef]
111. Song, T.; Fang, T.; Li, J.; Xu, L.; Zhang, F.; Han, B.; Shan, Q.; Zeng, H. Organic–Inorganic Hybrid Passivation Enables Perovskite QLEDs with an EQE of 16.48%. *Adv. Mater.* **2018**, *30*, 1805409. [CrossRef]
112. Swarnkar, A.; Marshall, A.R.; Sanehira, E.M.; Chernomordik, B.D.; Moore, D.T.; Christians, J.A.; Chakrabarti, T.; Luther, J.M. Quantum Dot–Induced Phase Stabilization of α-$CsPbI_3$ Perovskite for High-Efficiency Photovoltaics. *Science* **2016**, *354*, 92–95. [CrossRef] [PubMed]
113. Nedelcu, G.; Protesescu, L.; Yakunin, S.; Bodnarchuk, M.I.; Grotevent, M.J.; Kovalenko, M.V. Fast Anion-Exchange in Highly Luminescent Nanocrystals of Cesium Lead Halide Perovskites ($CsPbX_3$, X = Cl, Br, I). *Nano Lett.* **2015**, *15*, 5635–5640. [CrossRef] [PubMed]
114. Jia, D.; Chen, J.; Mei, X.; Fan, W.; Luo, S.; Yu, M.; Liu, J.; Zhang, X. Surface Matrix Curing of Inorganic $CsPbI_3$ Perovskite Quantum Dots for Solar Cells with Efficiency over 16%. *Energy Environ. Sci.* **2021**, *14*, 4599–4609. [CrossRef]
115. Lignos, I.; Stavrakis, S.; Nedelcu, G.; Protesescu, L.; deMello, A.J.; Kovalenko, M.V. Synthesis of Cesium Lead Halide Perovskite Nanocrystals in a Droplet-Based Microfluidic Platform: Fast Parametric Space Mapping. *Nano Lett.* **2016**, *16*, 1869–1877. [CrossRef]
116. Krieg, F.; Ochsenbein, S.T.; Yakunin, S.; ten Brinck, S.; Aellen, P.; Süess, A.; Clerc, B.; Guggisberg, D.; Nazarenko, O.; Shynkarenko, Y.; et al. Colloidal $CsPbX_3$ (X = Cl, Br, I) Nanocrystals 2.0: Zwitterionic Capping Ligands for Improved Durability and Stability. *ACS Energy Lett.* **2018**, *3*, 641–646. [CrossRef]
117. Shynkarenko, Y.; Bodnarchuk, M.I.; Bernasconi, C.; Berezovska, Y.; Verteletskyi, V.; Ochsenbein, S.T.; Kovalenko, M.V. Direct Synthesis of Quaternary Alkylammonium-Capped Perovskite Nanocrystals for Efficient Blue and Green Light-Emitting Diodes. *ACS Energy Lett.* **2019**, *4*, 2703–2711. [CrossRef]
118. Li, G.; Huang, J.; Zhu, H.; Li, Y.; Tang, J.-X.; Jiang, Y. Surface Ligand Engineering for Near-Unity Quantum Yield Inorganic Halide Perovskite QDs and High-Performance QLEDs. *Chem. Mater.* **2018**, *30*, 6099–6107. [CrossRef]
119. Dai, J.; Xi, J.; Li, L.; Zhao, J.; Shi, Y.; Zhang, W.; Ran, C.; Jiao, B.; Hou, X.; Duan, X.; et al. Charge Transport between Coupling Colloidal Perovskite Quantum Dots Assisted by Functional Conjugated Ligands. *Angew. Chem. Int. Ed.* **2018**, *57*, 5754–5758. [CrossRef]
120. Liu, F.; Zhang, Y.; Ding, C.; Kobayashi, S.; Izuishi, T.; Nakazawa, N.; Toyoda, T.; Ohta, T.; Hayase, S.; Minemoto, T.; et al. Highly Luminescent Phase-Stable $CsPbI_3$ Perovskite Quantum Dots Achieving Near 100% Absolute Photoluminescence Quantum Yield. *ACS Nano* **2017**, *11*, 10373–10383. [CrossRef]

121. Hao, M.; Bai, Y.; Zeiske, S.; Ren, L.; Liu, J.; Yuan, Y.; Zarrabi, N.; Cheng, N.; Ghasemi, M.; Chen, P.; et al. Ligand-Assisted Cation-Exchange Engineering for High-Efficiency Colloidal $Cs_{1-x}FA_xPbI_3$ Quantum Dot Solar Cells with Reduced Phase Segregation. *Nat. Energy* **2020**, *5*, 79–88. [CrossRef]
122. Wheeler, L.M.; Sanehira, E.M.; Marshall, A.R.; Schulz, P.; Suri, M.; Anderson, N.C.; Christians, J.A.; Nordlund, D.; Sokaras, D.; Kroll, T.; et al. Targeted Ligand-Exchange Chemistry on Cesium Lead Halide Perovskite Quantum Dots for High-Efficiency Photovoltaics. *J. Am. Chem. Soc.* **2018**, *140*, 10504–10513. [CrossRef] [PubMed]
123. Chen, J.; Jia, D.; Qiu, J.; Zhuang, R.; Hua, Y.; Zhang, X. Multidentate Passivation Crosslinking Perovskite Quantum Dots for Efficient Solar Cells. *Nano Energy* **2022**, *96*, 107140. [CrossRef]
124. Ling, X.; Zhou, S.; Yuan, J.; Shi, J.; Qian, Y.; Larson, B.W.; Zhao, Q.; Qin, C.; Li, F.; Shi, G.; et al. 14.1% $CsPbI_3$ Perovskite Quantum Dot Solar Cells via Cesium Cation Passivation. *Adv. Energy Mater.* **2019**, *9*, 1900721. [CrossRef]
125. Ling, X.; Yuan, J.; Zhang, X.; Qian, Y.; Zakeeruddin, S.M.; Larson, B.W.; Zhao, Q.; Shi, J.; Yang, J.; Ji, K.; et al. Guanidinium-Assisted Surface Matrix Engineering for Highly Efficient Perovskite Quantum Dot Photovoltaics. *Adv. Mater.* **2020**, *32*, 2001906. [CrossRef] [PubMed]
126. Kim, J.; Cho, S.; Dinic, F.; Choi, J.; Choi, C.; Jeong, S.M.; Lee, J.-S.; Voznyy, O.; Ko, M.J.; Kim, Y. Hydrophobic Stabilizer-Anchored Fully Inorganic Perovskite Quantum Dots Enhance Moisture Resistance and Photovoltaic Performance. *Nano Energy* **2020**, *75*, 104985. [CrossRef]
127. Wi, S.; Kim, H.; Chen, M.; Nam, H.; Guo, L.J.; Meyhofer, E.; Liang, X. Enhancement of Photovoltaic Response in Multilayer MoS_2 Induced by Plasma Doping. *ACS Nano* **2014**, *8*, 5270–5281. [CrossRef] [PubMed]
128. Guo, Q.; Pospischil, A.; Bhuiyan, M.; Jiang, H.; Tian, H.; Farmer, D.; Deng, B.; Li, C.; Han, S.-J.; Wang, H.; et al. Black Phosphorus Mid-Infrared Photodetectors with High Gain. *Nano Lett.* **2016**, *16*, 4648–4655. [CrossRef]
129. McVay, E.; Zubair, A.; Lin, Y.; Nourbakhsh, A.; Palacios, T. Impact of Al_2O_3 Passivation on the Photovoltaic Performance of Vertical WSe_2 Schottky Junction Solar Cells. *ACS Appl. Mater. Interfaces* **2020**, *12*, 57987–57995. [CrossRef]
130. Nassiri Nazif, K.; Daus, A.; Hong, J.; Lee, N.; Vaziri, S.; Kumar, A.; Nitta, F.; Chen, M.E.; Kananian, S.; Islam, R.; et al. High-Specific-Power Flexible Transition Metal Dichalcogenide Solar Cells. *Nat. Commun.* **2021**, *12*, 7034. [CrossRef]
131. Wang, X.; Wang, P.; Wang, J.; Hu, W.; Zhou, X.; Guo, N.; Huang, H.; Sun, S.; Shen, H.; Lin, T.; et al. Ultrasensitive and Broadband MoS_2 Photodetector Driven by Ferroelectrics. *Adv. Mater.* **2015**, *27*, 6575–6581. [CrossRef]
132. Kim, S.J.; Kim, D.; Min, B.K.; Yi, Y.; Mondal, S.; Nguyen, V.; Hwang, J.; Suh, D.; Cho, K.; Choi, C. Bandgap Tuned WS_2 Thin-Film Photodetector by Strain Gradient in van Der Waals Effective Homojunctions. *Adv. Opt. Mater.* **2021**, *9*, 2101310. [CrossRef]
133. Gan, X.; Shiue, R.-J.; Gao, Y.; Meric, I.; Heinz, T.F.; Shepard, K.; Hone, J.; Assefa, S.; Englund, D. Chip-Integrated Ultrafast Graphene Photodetector with High Responsivity. *Nat. Photonics* **2013**, *7*, 883–887. [CrossRef]
134. Kim, C.O.; Kim, S.; Shin, D.H.; Kang, S.S.; Kim, J.M.; Jang, C.W.; Joo, S.S.; Lee, J.S.; Kim, J.H.; Choi, S.-H.; et al. High Photoresponsivity in an All-Graphene p–n Vertical Junction Photodetector. *Nat. Commun.* **2014**, *5*, 3249. [CrossRef]
135. Schuler, S.; Schall, D.; Neumaier, D.; Schwarz, B.; Watanabe, K.; Taniguchi, T.; Mueller, T. Graphene Photodetector Integrated on a Photonic Crystal Defect Waveguide. *ACS Photonics* **2018**, *5*, 4758–4763. [CrossRef]
136. Ma, P.; Salamin, Y.; Baeuerle, B.; Josten, A.; Heni, W.; Emboras, A.; Leuthold, J. Plasmonically Enhanced Graphene Photodetector Featuring 100 Gbit/s Data Reception, High Responsivity, and Compact Size. *ACS Photonics* **2019**, *6*, 154–161. [CrossRef]
137. AlAloul, M.; Rasras, M. Plasmon-Enhanced Graphene Photodetector with CMOS-Compatible Titanium Nitride. *J. Opt. Soc. Am. B* **2021**, *38*, 602. [CrossRef]
138. Sulas-Kern, D.B.; Miller, E.M.; Blackburn, J.L. Photoinduced Charge Transfer in Transition Metal Dichalcogenide Heterojunctions—Towards next Generation Energy Technologies. *Energy Environ. Sci.* **2020**, *13*, 2684–2740. [CrossRef]
139. Ma, Y.; Dai, Y.; Guo, M.; Niu, C.; Lu, J.; Huang, B. Electronic and Magnetic Properties of Perfect, Vacancy-Doped, and Nonmetal Adsorbed $MoSe_2$, $MoTe_2$ and WS_2 Monolayers. *Phys. Chem. Chem. Phys.* **2011**, *13*, 15546. [CrossRef]
140. Fathipour, S.; Ma, N.; Hwang, W.S.; Protasenko, V.; Vishwanath, S.; Xing, H.G.; Xu, H.; Jena, D.; Appenzeller, J.; Seabaugh, A. Exfoliated Multilayer $MoTe_2$ Field-Effect Transistors. *Appl. Phys. Lett.* **2014**, *105*, 192101. [CrossRef]
141. Huang, H.; Wang, J.; Hu, W.; Liao, L.; Wang, P.; Wang, X.; Gong, F.; Chen, Y.; Wu, G.; Luo, W.; et al. Highly Sensitive Visible to Infrared $MoTe_2$ Photodetectors Enhanced by the Photogating Effect. *Nanotechnology* **2016**, *27*, 445201. [CrossRef]
142. Tran, V.; Soklaski, R.; Liang, Y.; Yang, L. Layer-Controlled Band Gap and Anisotropic Excitons in Few-Layer Black Phosphorus. *Phys. Rev. B* **2014**, *89*, 235319. [CrossRef]
143. Jalaei, S.; Karamdel, J.; Ghalami-Bavil-Olyaee, H. Mid-Infrared Photodetector Based on Selenium-Doped Black Phosphorus. *Phys. Status Solidi A* **2020**, *217*, 2000483. [CrossRef]
144. Fu, Q.; Wang, X.; Liu, F.; Dong, Y.; Liu, Z.; Zheng, S.; Chaturvedi, A.; Zhou, J.; Hu, P.; Zhu, Z.; et al. Ultrathin Ruddlesden–Popper Perovskite Heterojunction for Sensitive Photodetection. *Small* **2019**, *15*, 1902890. [CrossRef] [PubMed]
145. Kim, H.-S.; Lee, C.-R.; Im, J.-H.; Lee, K.-B.; Moehl, T.; Marchioro, A.; Moon, S.-J.; Humphry-Baker, R.; Yum, J.-H.; Moser, J.E.; et al. Lead Iodide Perovskite Sensitized All-Solid-State Submicron Thin Film Mesoscopic Solar Cell with Efficiency Exceeding 9%. *Sci. Rep.* **2012**, *2*, 591. [CrossRef] [PubMed]
146. Lee, M.M.; Teuscher, J.; Miyasaka, T.; Murakami, T.N.; Snaith, H.J. Efficient Hybrid Solar Cells Based on Meso-Superstructured Organometal Halide Perovskites. *Science* **2012**, *338*, 643–647. [CrossRef]
147. Burschka, J.; Pellet, N.; Moon, S.-J.; Humphry-Baker, R.; Gao, P.; Nazeeruddin, M.K.; Grätzel, M. Sequential Deposition as a Route to High-Performance Perovskite-Sensitized Solar Cells. *Nature* **2013**, *499*, 316–319. [CrossRef]

148. Yang, W.S.; Noh, J.H.; Jeon, N.J.; Kim, Y.C.; Ryu, S.; Seo, J.; Seok, S.I. High-Performance Photovoltaic Perovskite Layers Fabricated through Intramolecular Exchange. *Science* **2015**, *348*, 1234–1237. [CrossRef]
149. Bi, C.; Wang, Q.; Shao, Y.; Yuan, Y.; Xiao, Z.; Huang, J. Non-Wetting Surface-Driven High-Aspect-Ratio Crystalline Grain Growth for Efficient Hybrid Perovskite Solar Cells. *Nat. Commun.* **2015**, *6*, 7747. [CrossRef]
150. Serpetzoglou, E.; Konidakis, I.; Kakavelakis, G.; Maksudov, T.; Kymakis, E.; Stratakis, E. Improved Carrier Transport in Perovskite Solar Cells Probed by Femtosecond Transient Absorption Spectroscopy. *ACS Appl. Mater. Interfaces* **2017**, *9*, 43910–43919. [CrossRef]
151. Li, F.; Zhu, W.; Bao, C.; Yu, T.; Wang, Y.; Zhou, X.; Zou, Z. Laser-Assisted Crystallization of $CH_3NH_3PbI_3$ Films for Efficient Perovskite Solar Cells with a High Open-Circuit Voltage. *Chem. Commun.* **2016**, *52*, 5394–5397. [CrossRef]
152. Jeon, T.; Jin, H.M.; Lee, S.H.; Lee, J.M.; Park, H.I.; Kim, M.K.; Lee, K.J.; Shin, B.; Kim, S.O. Laser Crystallization of Organic–Inorganic Hybrid Perovskite Solar Cells. *ACS Nano* **2016**, *10*, 7907–7914. [CrossRef]
153. Konidakis, I.; Maksudov, T.; Serpetzoglou, E.; Kakavelakis, G.; Kymakis, E.; Stratakis, E. Improved Charge Carrier Dynamics of $CH_3NH_3PbI_3$ Perovskite Films Synthesized by Means of Laser-Assisted Crystallization. *ACS Appl. Energy Mater.* **2018**, *1*, 5101–5111. [CrossRef]
154. Jung, E.H.; Jeon, N.J.; Park, E.Y.; Moon, C.S.; Shin, T.J.; Yang, T.-Y.; Noh, J.H.; Seo, J. Efficient, Stable and Scalable Perovskite Solar Cells Using Poly(3-Hexylthiophene). *Nature* **2019**, *567*, 511–515. [CrossRef]
155. Yan, J.; Qiu, W.; Wu, G.; Heremans, P.; Chen, H. Recent Progress in 2D/Quasi-2D Layered Metal Halide Perovskites for Solar Cells. *J. Mater. Chem. A* **2018**, *6*, 11063–11077. [CrossRef]
156. Liu, P.; Han, N.; Wang, W.; Ran, R.; Zhou, W.; Shao, Z. High-Quality Ruddlesden–Popper Perovskite Film Formation for High-Performance Perovskite Solar Cells. *Adv. Mater.* **2021**, *33*, 2002582. [CrossRef] [PubMed]
157. Shao, S.; Duim, H.; Wang, Q.; Xu, B.; Dong, J.; Adjokatse, S.; Blake, G.R.; Protesescu, L.; Portale, G.; Hou, J.; et al. Tuning the Energetic Landscape of Ruddlesden–Popper Perovskite Films for Efficient Solar Cells. *ACS Energy Lett.* **2020**, *5*, 39–46. [CrossRef]
158. Yang, R.; Li, R.; Cao, Y.; Wei, Y.; Miao, Y.; Tan, W.L.; Jiao, X.; Chen, H.; Zhang, L.; Chen, Q.; et al. Oriented Quasi-2D Perovskites for High Performance Optoelectronic Devices. *Adv. Mater.* **2018**, *30*, 1804771. [CrossRef]
159. Chen, K.; Jin, W.; Zhang, Y.; Yang, T.; Reiss, P.; Zhong, Q.; Bach, U.; Li, Q.; Wang, Y.; Zhang, H.; et al. High Efficiency Mesoscopic Solar Cells Using $CsPbI_3$ Perovskite Quantum Dots Enabled by Chemical Interface Engineering. *J. Am. Chem. Soc.* **2020**, *142*, 3775–3783. [CrossRef]
160. Yuan, J.; Hazarika, A.; Zhao, Q.; Ling, X.; Moot, T.; Ma, W.; Luther, J.M. Metal Halide Perovskites in Quantum Dot Solar Cells: Progress and Prospects. *Joule* **2020**, *4*, 1160–1185. [CrossRef]
161. Bao, C.; Yang, J.; Bai, S.; Xu, W.; Yan, Z.; Xu, Q.; Liu, J.; Zhang, W.; Gao, F. High Performance and Stable All-Inorganic Metal Halide Perovskite-Based Photodetectors for Optical Communication Applications. *Adv. Mater.* **2018**, *30*, 1803422. [CrossRef]
162. Leung, S.-F.; Ho, K.-T.; Kung, P.-K.; Hsiao, V.K.S.; Alshareef, H.N.; Wang, Z.L.; He, J.-H. A Self-Powered and Flexible Organometallic Halide Perovskite Photodetector with Very High Detectivity. *Adv. Mater.* **2018**, *30*, 1704611. [CrossRef]
163. Xu, Z.; Li, Y.; Liu, X.; Ji, C.; Chen, H.; Li, L.; Han, S.; Hong, M.; Luo, J.; Sun, Z. Highly Sensitive and Ultrafast Responding Array Photodetector Based on a Newly Tailored 2D Lead Iodide Perovskite Crystal. *Adv. Opt. Mater.* **2019**, *7*, 1900308. [CrossRef]
164. Loi, H.; Cao, J.; Guo, X.; Liu, C.; Wang, N.; Song, J.; Tang, G.; Zhu, Y.; Yan, F. Gradient 2D/3D Perovskite Films Prepared by Hot-Casting for Sensitive Photodetectors. *Adv. Sci.* **2020**, *7*, 2000776. [CrossRef] [PubMed]
165. Ma, X.; Xu, Y.; Li, S.; Lo, T.W.; Zhang, B.; Rogach, A.L.; Lei, D. A Flexible Plasmonic-Membrane-Enhanced Broadband Tin-Based Perovskite Photodetector. *Nano Lett.* **2021**, *21*, 9195–9202. [CrossRef]
166. Qian, L.; Sun, Y.; Sun, M.; Fang, Z.; Li, L.; Xie, D.; Li, C.; Ding, L. 2D Perovskite Microsheets for High-Performance Photodetectors. *J. Mater. Chem. C* **2019**, *7*, 5353–5358. [CrossRef]
167. Qian, L.; Sun, Y.; Wu, M.; Li, C.; Xie, D.; Ding, L.; Shi, G. A Lead-Free Two-Dimensional Perovskite for a High-Performance Flexible Photoconductor and a Light-Stimulated Synaptic Device. *Nanoscale* **2018**, *10*, 6837–6843. [CrossRef] [PubMed]
168. Zhang, X.; Wang, Q.; Jin, Z.; Zhang, J.; Liu, S. (Frank) Stable Ultra-Fast Broad-Bandwidth Photodetectors Based on α-$CsPbI_3$ Perovskite and $NaYF_4$:Yb,Er Quantum Dots. *Nanoscale* **2017**, *9*, 6278–6285. [CrossRef] [PubMed]
169. Chen, H.; Hou, Y.; Halbig, C.E.; Chen, S.; Zhang, H.; Li, N.; Guo, F.; Tang, X.; Gasparini, N.; Levchuk, I.; et al. Extending the Environmental Lifetime of Unpackaged Perovskite Solar Cells through Interfacial Design. *J. Mater. Chem. A* **2016**, *4*, 11604–11610. [CrossRef]
170. Singh, R.; Giri, A.; Pal, M.; Thiyagarajan, K.; Kwak, J.; Lee, J.-J.; Jeong, U.; Cho, K. Perovskite Solar Cells with an MoS_2 Electron Transport Layer. *J. Mater. Chem. A* **2019**, *7*, 7151–7158. [CrossRef]
171. Sherkar, T.S.; Momblona, C.; Gil-Escrig, L.; Ávila, J.; Sessolo, M.; Bolink, H.J.; Koster, L.J.A. Recombination in Perovskite Solar Cells: Significance of Grain Boundaries, Interface Traps, and Defect Ions. *ACS Energy Lett.* **2017**, *2*, 1214–1222. [CrossRef]
172. Wang, Y.; Hu, Y.; Han, D.; Yuan, Q.; Cao, T.; Chen, N.; Zhou, D.; Cong, H.; Feng, L. Ammonia-Treated Graphene Oxide and PEDOT:PSS as Hole Transport Layer for High-Performance Perovskite Solar Cells with Enhanced Stability. *Org. Electron.* **2019**, *70*, 63–70. [CrossRef]
173. Yun, J.-M.; Yeo, J.-S.; Kim, J.; Jeong, H.-G.; Kim, D.-Y.; Noh, Y.-J.; Kim, S.-S.; Ku, B.-C.; Na, S.-I. Solution-Processable Reduced Graphene Oxide as a Novel Alternative to PEDOT:PSS Hole Transport Layers for Highly Efficient and Stable Polymer Solar Cells. *Adv. Mater.* **2011**, *23*, 4923–4928. [CrossRef] [PubMed]

74. Yeo, J.-S.; Kang, R.; Lee, S.; Jeon, Y.-J.; Myoung, N.; Lee, C.-L.; Kim, D.-Y.; Yun, J.-M.; Seo, Y.-H.; Kim, S.-S.; et al. Highly Efficient and Stable Planar Perovskite Solar Cells with Reduced Graphene Oxide Nanosheets as Electrode Interlayer. *Nano Energy* **2015**, *12*, 96–104. [CrossRef]
75. Jokar, E.; Huang, Z.Y.; Narra, S.; Wang, C.-Y.; Kattoor, V.; Chung, C.-C.; Diau, E.W.-G. Anomalous Charge-Extraction Behavior for Graphene-Oxide (GO) and Reduced Graphene-Oxide (RGO) Films as Efficient p-Contact Layers for High-Performance Perovskite Solar Cells. *Adv. Energy Mater.* **2018**, *8*, 1701640. [CrossRef]
76. Agresti, A.; Pescetelli, S.; Cinà, L.; Konios, D.; Kakavelakis, G.; Kymakis, E.; Carlo, A.D. Efficiency and Stability Enhancement in Perovskite Solar Cells by Inserting Lithium-Neutralized Graphene Oxide as Electron Transporting Layer. *Adv. Funct. Mater.* **2016**, *26*, 2686–2694. [CrossRef]
77. Batmunkh, M.; Vimalanathan, K.; Wu, C.; Bati, A.S.R.; Yu, L.; Tawfik, S.A.; Ford, M.J.; Macdonald, T.J.; Raston, C.L.; Priya, S.; et al. Efficient Production of Phosphorene Nanosheets via Shear Stress Mediated Exfoliation for Low-Temperature Perovskite Solar Cells. *Small Methods* **2019**, *3*, 1800521. [CrossRef]
78. Zhao, X.; Liu, S.; Zhang, H.; Chang, S.-Y.; Huang, W.; Zhu, B.; Shen, Y.; Shen, C.; Wang, D.; Yang, Y.; et al. 20% Efficient Perovskite Solar Cells with 2D Electron Transporting Layer. *Adv. Funct. Mater.* **2019**, *29*, 1805168. [CrossRef]
79. Kim, Y.G.; Kwon, K.C.; Le, Q.V.; Hong, K.; Jang, H.W.; Kim, S.Y. Atomically Thin Two-Dimensional Materials as Hole Extraction Layers in Organolead Halide Perovskite Photovoltaic Cells. *J. Power Sources* **2016**, *319*, 1–8. [CrossRef]
80. Chen, L.-C.; Tseng, Z.-L.; Chen, C.-C.; Chang, S.H.; Ho, C.-H. Fabrication and Characteristics of $CH_3NH_3PbI_3$ Perovskite Solar Cells with Molybdenum-Selenide Hole-Transport Layer. *Appl. Phys. Express* **2016**, *9*, 122301. [CrossRef]
81. Yang, L.; Dall'Agnese, C.; Dall'Agnese, Y.; Chen, G.; Gao, Y.; Sanehira, Y.; Jena, A.K.; Wang, X.; Gogotsi, Y.; Miyasaka, T. Surface-Modified Metallic $Ti_3C_2T_x$ MXene as Electron Transport Layer for Planar Heterojunction Perovskite Solar Cells. *Adv. Funct. Mater.* **2019**, *29*, 1905694. [CrossRef]
82. Furchi, M.; Urich, A.; Pospischil, A.; Lilley, G.; Unterrainer, K.; Detz, H.; Klang, P.; Andrews, A.M.; Schrenk, W.; Strasser, G.; et al. Microcavity-Integrated Graphene Photodetector. *Nano Lett.* **2012**, *12*, 2773–2777. [CrossRef] [PubMed]
83. Konstantatos, G.; Badioli, M.; Gaudreau, L.; Osmond, J.; Bernechea, M.; de Arquer, F.P.G.; Gatti, F.; Koppens, F.H.L. Hybrid Graphene–Quantum Dot Phototransistors with Ultrahigh Gain. *Nat. Nanotechnol.* **2012**, *7*, 363–368. [CrossRef] [PubMed]
84. Lee, Y.; Kwon, J.; Hwang, E.; Ra, C.-H.; Yoo, W.J.; Ahn, J.-H.; Park, J.H.; Cho, J.H. High-Performance Perovskite-Graphene Hybrid Photodetector. *Adv. Mater.* **2015**, *27*, 41–46. [CrossRef] [PubMed]
85. Zou, Y.; Zou, T.; Zhao, C.; Wang, B.; Xing, J.; Yu, Z.; Cheng, J.; Xin, W.; Yang, J.; Yu, W.; et al. A Highly Sensitive Single Crystal Perovskite–Graphene Hybrid Vertical Photodetector. *Small* **2020**, *16*, 2000733. [CrossRef]
86. Bera, K.P.; Haider, G.; Huang, Y.-T.; Roy, P.K.; Paul Inbaraj, C.R.; Liao, Y.-M.; Lin, H.-I.; Lu, C.-H.; Shen, C.; Shih, W.Y.; et al. Graphene Sandwich Stable Perovskite Quantum-Dot Light-Emissive Ultrasensitive and Ultrafast Broadband Vertical Phototransistors. *ACS Nano* **2019**, *13*, 12540–12552. [CrossRef]
87. Chen, Z.; Kang, Z.; Rao, C.; Cheng, Y.; Liu, N.; Zhang, Z.; Li, L.; Gao, Y. Improving Performance of Hybrid Graphene–Perovskite Photodetector by a Scratch Channel. *Adv. Electron. Mater.* **2019**, *5*, 1900168. [CrossRef]
88. Lu, J.; Carvalho, A.; Liu, H.; Lim, S.X.; Castro Neto, A.H.; Sow, C.H. Hybrid Bilayer WSe_2-$CH_3NH_3PbI_3$ Organolead Halide Perovskite as a High-Performance Photodetector. *Angew. Chem.* **2016**, *128*, 12124–12128. [CrossRef]
89. Wang, Y.; Fullon, R.; Acerce, M.; Petoukhoff, C.E.; Yang, J.; Chen, C.; Du, S.; Lai, S.K.; Lau, S.P.; Voiry, D.; et al. Solution-Processed MoS_2/Organolead Trihalide Perovskite Photodetectors. *Adv. Mater.* **2017**, *29*, 1603995. [CrossRef]
190. Erkılıç, U.; Solís-Fernández, P.; Ji, H.G.; Shinokita, K.; Lin, Y.-C.; Maruyama, M.; Suenaga, K.; Okada, S.; Matsuda, K.; Ago, H. Vapor Phase Selective Growth of Two-Dimensional Perovskite/WS_2 Heterostructures for Optoelectronic Applications. *ACS Appl. Mater. Interfaces* **2019**, *11*, 40503–40511. [CrossRef]
191. Wu, H.; Si, H.; Zhang, Z.; Kang, Z.; Wu, P.; Zhou, L.; Zhang, S.; Zhang, Z.; Liao, Q.; Zhang, Y. All-Inorganic Perovskite Quantum Dot-Monolayer MoS_2 Mixed-Dimensional van Der Waals Heterostructure for Ultrasensitive Photodetector. *Adv. Sci.* **2018**, *5*, 1801219. [CrossRef]
192. Wang, Y.; Zou, X.; Lin, J.; Jiang, J.; Liu, Y.; Liu, X.; Zhao, X.; Liu, Y.F.; Ho, J.C.; Liao, L. Perovskite/Black Phosphorus/MoS_2 Photogate Reversed Photodiodes with Ultrahigh Light On/Off Ratio and Fast Response. *ACS Nano* **2019**, *13*, 4804–4813. [CrossRef] [PubMed]
193. Muduli, S.; Pandey, P.; Devatha, G.; Babar, R.; Thripuranthaka, M.; Kothari, D.C.; Kabir, M.; Pillai, P.P.; Ogale, S. Photoluminescence Quenching in Self-Assembled $CsPbBr_3$ Quantum Dots on Few-Layer Black Phosphorus Sheets. *Angew. Chem.* **2018**, *130*, 7808–7812. [CrossRef]
194. Deng, W.; Huang, H.; Jin, H.; Li, W.; Chu, X.; Xiong, D.; Yan, W.; Chun, F.; Xie, M.; Luo, C.; et al. All-Sprayed-Processable, Large-Area, and Flexible Perovskite/MXene-Based Photodetector Arrays for Photocommunication. *Adv. Opt. Mater.* **2019**, *7*, 1801521. [CrossRef]
195. Liang, C.; Zhao, D.; Li, Y.; Li, X.; Peng, S.; Shao, G.; Xing, G. Ruddlesden-Popper Perovskite for Stable Solar Cells. *Energy Environ. Mater.* **2018**, *1*, 221–231. [CrossRef]
196. Lin, Y.; Bai, Y.; Fang, Y.; Wang, Q.; Deng, Y.; Huang, J. Suppressed Ion Migration in Low-Dimensional Perovskites. *ACS Energy Lett.* **2017**, *2*, 1571–1572. [CrossRef]
197. Xu, Z.; Lu, D.; Dong, X.; Chen, M.; Fu, Q.; Liu, Y. Highly Efficient and Stable Dion–Jacobson Perovskite Solar Cells Enabled by Extended Π-Conjugation of Organic Spacer. *Adv. Mater.* **2021**, *33*, 2105083. [CrossRef]

198. Zhang, Q.; Wang, B.; Zheng, W.; Kong, L.; Wan, Q.; Zhang, C.; Li, Z.; Cao, X.; Liu, M.; Li, L. Ceramic-like Stable CsPbBr$_3$ Nanocrystals Encapsulated in Silica Derived from Molecular Sieve Templates. *Nat. Commun.* **2020**, *11*, 31. [CrossRef]
199. Tan, C.; Chen, J.; Wu, X.-J.; Zhang, H. Epitaxial Growth of Hybrid Nanostructures. *Nat. Rev. Mater.* **2018**, *3*, 17089. [CrossRef]
200. Huang, G.; Wang, C.; Xu, S.; Zong, S.; Lu, J.; Wang, Z.; Lu, C.; Cui, Y. Postsynthetic Doping of MnCl$_2$ Molecules into Preformed CsPbBr$_3$ Perovskite Nanocrystals via a Halide Exchange-Driven Cation Exchange. *Adv. Mater.* **2017**, *29*, 1700095. [CrossRef]
201. Zhang, L.; Kang, C.; Zhang, G.; Pan, Z.; Huang, Z.; Xu, S.; Rao, H.; Liu, H.; Wu, S.; Wu, X.; et al. All-Inorganic CsPbI$_3$ Quantum Dot Solar Cells with Efficiency over 16% by Defect Control. *Adv. Funct. Mater.* **2021**, *31*, 2005930. [CrossRef]
202. Liu, F.; Ding, C.; Zhang, Y.; Ripolles, T.S.; Kamisaka, T.; Toyoda, T.; Hayase, S.; Minemoto, T.; Yoshino, K.; Dai, S.; et al. Colloidal Synthesis of Air-Stable Alloyed CsSn$_{1-x}$Pb$_x$I$_3$ Perovskite Nanocrystals for Use in Solar Cells. *J. Am. Chem. Soc.* **2017**, *139*, 16708–16719. [CrossRef] [PubMed]

Article

Hierarchically Assembled Plasmonic Metal-Dielectric-Metal Hybrid Nano-Architectures for High-Sensitivity SERS Detection

Puran Pandey [1], Min-Kyu Seo [1], Ki Hoon Shin [1], Young-Woo Lee [2,*] and Jung Inn Sohn [1,*]

[1] Division of Physics and Semiconductor Science, Dongguk University-Seoul, Seoul 04620, Korea; ppcpurans@gmail.com (P.P.); seominkyuu@gmail.com (M.-K.S.); kihoonshin@dongguk.edu (K.H.S.)
[2] Department of Energy Systems, Soonchunhyang University, Asan-si 31538, Korea
* Correspondence: ywlee@sch.ac.kr (Y.-W.L.); junginn.sohn@dongguk.edu (J.I.S.)

Abstract: In this work, we designed and prepared a hierarchically assembled 3D plasmonic metal-dielectric-metal (PMDM) hybrid nano-architecture for high-performance surface-enhanced Raman scattering (SERS) sensing. The fabrication of the PMDM hybrid nanostructure was achieved by the thermal evaporation of Au film followed by thermal dewetting and the atomic layer deposition (ALD) of the Al_2O_3 dielectric layer, which is crucial for creating numerous nanogaps between the core Au and the out-layered Au nanoparticles (NPs). The PMDM hybrid nanostructures exhibited strong SERS signals originating from highly enhanced electromagnetic (EM) hot spots at the 3 nm Al_2O_3 layer serving as the nanogap spacer, as confirmed by the finite-difference time-domain (FDTD) simulation. The PMDM SERS substrate achieved an outstanding SERS performance, including a high sensitivity (enhancement factor, EF of 1.3×10^8 and low detection limit 10^{-11} M) and excellent reproducibility (relative standard deviation (RSD) < 7.5%) for rhodamine 6G (R6G). This study opens a promising route for constructing multilayered plasmonic structures with abundant EM hotspots for the highly sensitive, rapid, and reproducible detection of biomolecules.

Keywords: SERS; metal-dielectric-metal; Au nanoparticles; hot spots; FDTD simulation

Citation: Pandey, P.; Seo, M.-K.; Shin, K.H.; Lee, Y.-W.; Sohn, J.I. Hierarchically Assembled Plasmonic Metal-Dielectric-Metal Hybrid Nano-Architectures for High-Sensitivity SERS Detection. *Nanomaterials* **2022**, *12*, 401. https://doi.org/10.3390/nano12030401

Academic Editor: Onofrio M. Maragò

Received: 4 January 2022
Accepted: 24 January 2022
Published: 26 January 2022

Publisher's Note: MDPI stays neutral with regard to jurisdictional claims in published maps and institutional affiliations.

Copyright: © 2022 by the authors. Licensee MDPI, Basel, Switzerland. This article is an open access article distributed under the terms and conditions of the Creative Commons Attribution (CC BY) license (https://creativecommons.org/licenses/by/4.0/).

1. Introduction

Owing to its extremely high sensitivity, ability to work in real time, and multiplexing detection capability, surface-enhanced Raman scattering (SERS) has emerged as a powerful detection technique for sensing molecules through its unique fingerprint vibrational spectrum [1–5]. It has tremendous potential for single-molecule level detection [6,7], the investigation of live cells [8,9], the monitoring of catalytic reactions [10,11], and sensing molecules, in both liquid and solid samples [12,13]. In SERS, the Raman signals of analytes can be amplified by several orders of magnitude (10^8–10^{10}) based on two mechanisms: electromagnetic mechanism (enhancement of ~10^6–10^8) and chemical mechanism (enhancement of ~10^2–10^4) [14–16]. The SERS enhancement mostly relies on the amplification of the electromagnetic field—i.e., electromagnetic (EM) hot spots generated by the excitation of the localized surface plasmon resonance (LSPR) of the metal nanostructures [17]. Therefore, plasmonic metallic nanostructures including Au, Ag, and Cu have been fabricated to prepare an excellent SERS substrate for sensing molecules [18–20]. Designing and optimizing the geometry of a plasmonic nanostructure, such as its size [21], sharp corners [22], tips [23], surface roughness [24], and interparticle gaps [25], is essential in order to provide a strong EM hotspot and hence enhance SERS signal intensities. Among these, the interparticle gap structure has attracted considerable attention thanks to its ability to provide extremely strong EM hot spots within a sub-nanometer gap [26]. The precise control of nanogaps between plasmonic nanoparticles (NPs) at a nanometer scale is crucial to produce a high density of strong and stable EM hot spots. To maintain the specific sub-nanometer gap, a dielectric layer can be considered as a nanogap spacer between two layered plasmonic metal nanostructures—namely, metal-dielectric-metal hybrid nano-architectures [27–30]. The dielectric spacer offers several benefits: protecting the plasmonic core from oxidation, tunning the LSPR properties, and maintaining a sub-nanometer gap between metal nanostructures to obtain a strong EM hotspot [31–34]. Therefore, it is of great significance to construct a unique 3D nano-architecture SERS substrate that comprises a hierarchical assembly of plasmonic NPs, separated by a dielectric spacer, for achieving an extremely high SERS activity.

Inspired by the above discussion, we report a facile method for fabricating abundant nanogaps containing hierarchically assembled 3D plasmonic metal-dielectric-metal (PMDM) hybrid nano-architectures for superior SERS detection in this work. However, the developed PMDM hybrid SERS sensor is considerably different from the above-mentioned metal-dielectric-metal structure in terms of the preparation method and architecture of the SERS platform. The PMDM hybrid nanostructures were prepared by the thermal evaporation of Au film followed by the thermal annealing and atomic layer deposition (ALD) of the Al_2O_3 dielectric layer. We achieved an enormous SERS enhancement of the PMDM hybrid nanostructures, with a maximum enhancement factor (EF) of 1.3×10^8 and a low detection limit of 10^{-11} M R6G molecules. We further observed the excellent reproducibility of the SERS substrate with relative standard deviation (RSD) values of less than 7.5%. To support the experimental SERS performance, we conducted the finite-difference time-domain (FDTD) simulation of hybrid nanostructures and showed that a high density of strong EM hot spots was produced between the Au core and numerous out-layered Au NPs at the Al_2O_3 nanogap regions.

2. Materials and Methods

2.1. Hybrid Nanostructures Fabrication

SiO_2/Si substrates (purchased from Sehyoung wafertech, Seoul, Korea) were cleaned by acetone, isopropyl alcohol (Sigma-Aldrich, Saint Louis, MO, USA), and deionized water for 10 mins in sequence in an ultrasonic cleaner and dried at room temperature. Au films (10 nm) were deposited on SiO_2/Si substrates using thermal evaporation, where the thickness of films was controlled by the deposition time. The 10 nm Au films was annealed in a rapid thermal annealing (RTA) chamber at 800 °C for 120 s for the fabrication of core Au NPs arrays based on the solid-state dewetting [35,36]. Subsequently, the ultrathin dielectric layer—i.e., 3 nm Al_2O_3 film—was deposited on the as-prepared core Au NPs via the ALD method. Then, the 5nm Au films were deposited on the surface of Au/Al_2O_3 nanostructures, followed by annealing at 450 °C for 120 s to produce highly dense small-sized Au NPs on the surface of Au/Al_2O_3 NPs, which are referred to as a hierarchically assembled PMDM (Au/Al_2O_3/Au) nanostructures. For comparison, double dewetting Au nanostructures were prepared by a similar method without using an Al_2O_3 spacer.

2.2. Sample Characterization

The morphological characterization and elemental analysis of the as-synthesized hybrid nanostructures fabricated on SiO_2/Si substrates was performed using a field emission scanning electron microscope (FE-SEM, Hitachi S-7400, Tokyo, Japan), coupled with the energy-dispersive x-ray spectroscopy (EDS) analysis. Moreover, the crystalline information was examined by X-ray diffraction (XRD, Rigaku Ultima IV diffractometer, Tokyo, Japan) with Cu-Kα radiation, whereas the chemical states were evaluated using X-ray photoelectron spectroscopy (XPS, Veresprobe II, Ulvac-phi, Chigasaki, Japan).

2.3. SERS Analysis

Rhodamine 6G (R6G, Sigma-Aldrich, Saint Louis, MO, USA) was used as a probe molecule to determine the SERS activity of the as-prepared PMDM hybrid nanostructures. For the preparation of the samples for SERS measurements, the SERS substrates were immersed in different concentrations of R6G solution ranging from 10^{-12} to 10^{-5} M for

2 h to allow the sufficient adsorption of R6G molecules on plasmonic nanostructures. SERS measurements were performed using confocal Raman spectroscopy (HEDA, NOST, Seongnam, Korea) at room temperature. SERS signals were acquired using an incident laser with a wavelength of 532 nm with a power of 0.1 mW (laser spot size ~1 µm), 100× objective lens (numerical aperture = 0.80), and acquisition time of 10 s. To determine the EF, the Raman spectrum of R6G (10^{-2} M) adsorbed on the SiO_2/Si substrates was evaluated as above.

2.4. FDTD Simulation

The EM field distribution was calculated with the FDTD method (Lumerical Solutions Inc., Vancouver, BC, Canada). In our simplified unit of the simulation model, the core Au NP was assumed to be a larger hemisphere coated with a dielectric spacer layer (3 nm Al_2O_3) followed by the out-layered small Au NPs. The diameter of the core, top, and surrounding Au NPs was supposed to be 180, 60, and 30 nm, respectively. Furthermore, the incident light source of a plane wave, surrounding medium (air), perfectly matched layer (PML) as an absorption boundary in z-boundary, periodic boundary condition for x and y directions, and mesh size (1 nm) were selected for the simulation to compute the EM field distribution. The near-field EM field intensity was calculated in the vicinity of the nanostructures using two monitors in X-Y and Y-Z directions. The data of a refractive index for Au were obtained from the Johnson and Christy model [37]. The data for SiO_2 and Al_2O_3 were acquired from the model data provided by the software.

3. Results and Discussion

Figure 1a shows the fabrication procedures and surface morphology of the hierarchically assembled PMDM hybrid nanostructures fabricated on Si/SiO_2 substrate. The combination of double dewetting and the ALD approach was employed for the preparation of the SERS substrates. The aim of using this approach with a dielectric layer between Au NPs is to obtain hierarchical nano-architectures with a strong plasmonic response and massive gap-introduced EM hot spots. First, high-density core Au NPs arrays were prepared based on the thermal dewetting of 10 nm Au thin film at 800 °C for 120 s. The average diameter Au NPs was found to be ~136 nm and the corresponding size distribution is shown in the histogram of Figure S1. The surface morphology of the fabricated Au NPs on the substrate with well-dispersed semispherical or somewhat faceted NPs is shown in the SEM image of Figure 1b. Next, we deposited 5 nm Au films on as-prepared core Au NPs arrays and then annealed them at 450 °C for 120 s. This repeated dewetting of Au films resulted in the formation of a very high density of Au NPs, as shown in Figure 1c. The larger core Au NPs were surrounded by comparatively small Au NPs. However, the spacing between them was too large, meaning they cannot be considered a good candidate for SERS substrates. Therefore, we deposited a 3 nm Al_2O_3 thin film on core Au NPs to fabricate a metal-dielectric core-shell nanostructure via the ALD approach. Subsequently, a 5 nm Au thin film was deposited on the Au/Al_2O_3 nanostructure followed by annealing at 450 °C for 120 s, which gives rise to the formation of hierarchically assembled PMDM hybrid nanostructures, as shown in Figure 1d. These hierarchical PMDM hybrid nanostructures provide not only an increased surface coverage and roughness, but also multiple out-layered small-sized Au NPs separated with a nanogap layer of dielectric Al_2O_3 from the core Au NPs. The out-layered Au NPs can be distinctly observed along the surface of core Au NPs in Figure 1d. Moreover, we confirmed the formation of PMDM hybrid nanostructures by using EDS mapping based on elemental analysis, as shown in Figure 1e–h.

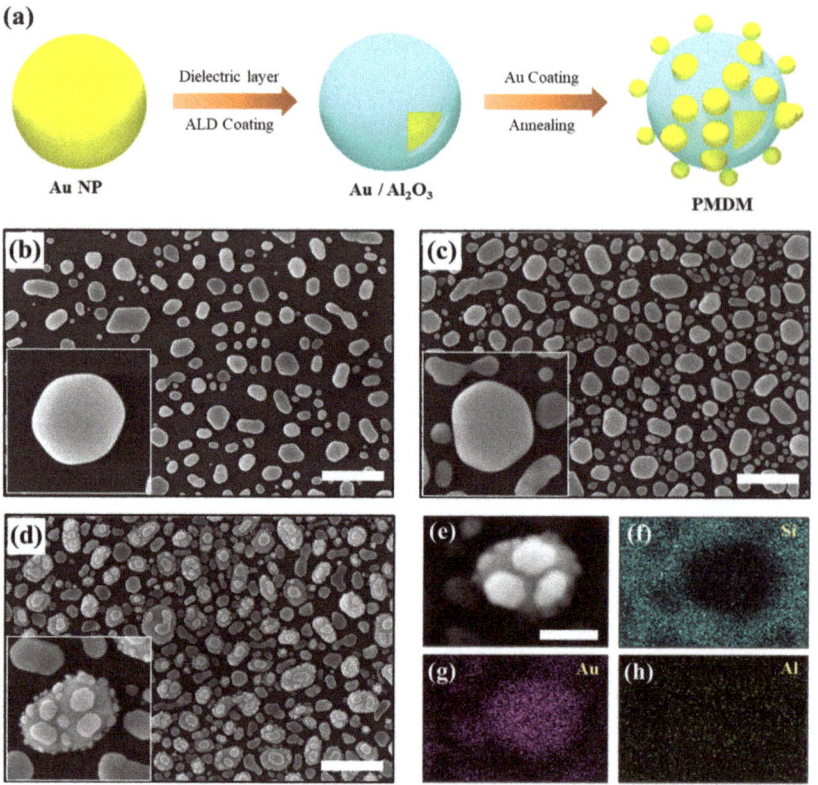

Figure 1. (**a**) Schematic illustration of the fabrication of hierarchically assembled plasmonic metal-dielectric-metal (PMDM) nano-architectures. (**b**) SEM image of Au nanoparticles (NPs) fabricated on Si/SiO$_2$ substrate by the annealing of Au film (10 nm) at 800 °C for 120 s. (**c**) Fabrication of double dewetting Au NPs arrays grown on Au NPs (Au/Au) by the deposition of 5 nm Au film and subsequent annealing at 450 °C for 120 s. (**d**) SEM image of hierarchically assembled PMDM hybrid nano-architectures. Scale bar: 400 nm. (**e**–**h**) EDS mapping of PMDM hybrid nanostructures grown on Si/SiO$_2$ substrate, where elemental maps are Si (blue), Au (red), and Al (green). Scale bar: 100 nm.

The crystalline structures of the Au and hybrid nanostructures were examined with an XRD pattern, in which all samples possessed almost the same diffraction peaks as those shown in Figure 2a. Four distinct diffraction peaks were observed at 38.3, 44.3, 64.7, and 77.7° corresponding to the (111), (200), (220), and (311) planes of the face-centered cubic phase of Au (JCPDS no. 04-784), revealing the formation of Au NPs. Furthermore, the XPS spectra of the Au and PMDM hybrid nanostructures were thoroughly analyzed to confirm the elemental and chemical states. Figure 2b shows the XPS survey spectra of the Au NPs, double dewetted Au/Au NPs, and PMDM hybrid nanostructures, discovering all the elements as expected. In particular, the Au 4f, Al 2p, Au 4d, and O 1s elements are all presented in the XPS survey spectra of PMDM hybrid nanostructures. As shown in the high-resolution XPS spectrum of Au 4f (Figure 2c), two characteristic peaks located at binding energies 84.2 and 87.9 eV are attributed to 4f$_{7/2}$ and 4f$_{5/2}$, respectively, indicating the presence of a metallic state of Au [38]. In addition, the high-resolution XPS analysis (Figure 2d) depicts the peak at 74.1 eV assigned to Al 2p, originating from the Al$_2$O$_3$ film [39]. The above evidence reveals the existence of dielectric spacer Al$_2$O$_3$ in the PMDM hybrid nanostructures.

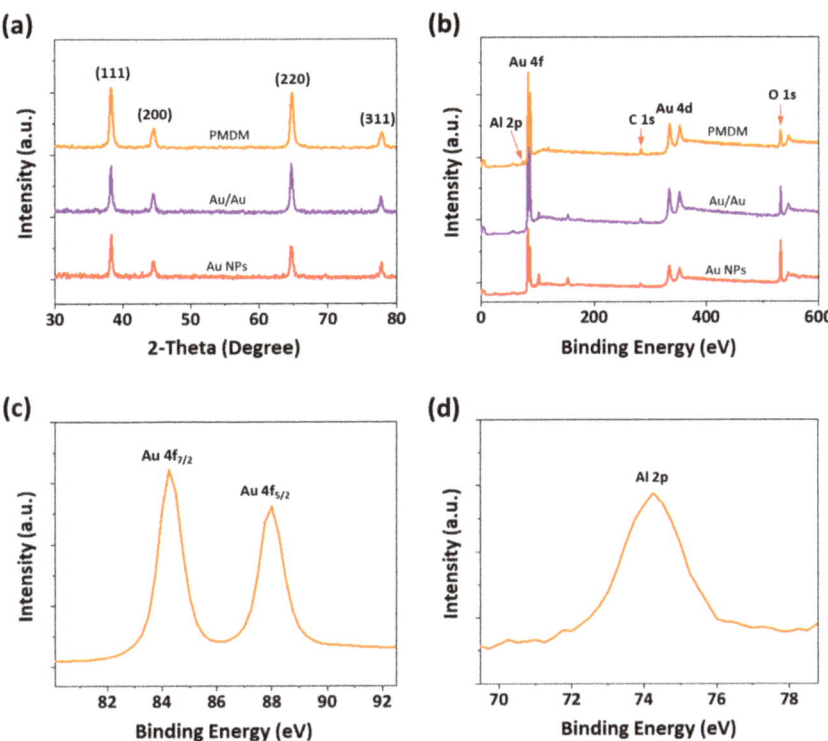

Figure 2. (a) XRD patterns of three different plasmonic nanostructures: Au, Au/Au, and PMDM hybrid nanostructures, as labeled. (b) XPS survey spectra of Au, Au/Au, and PMDM hybrid nanostructures. (c,d) High-resolution XPS spectra of Au 4f and Al 2p of PMDM hybrid nanostructure.

Next, FDTD simulations were used to calculate the spatial distribution of the near-field EM field of the plasmonic Au and PMDM hybrid nanostructures deposited on SiO_2/Si substrate. The FDTD simulation models were constructed by mimicking the real experimental results of nanostructures obtained from SEM images, as shown in Figure 3a,d. The FDTD simulation of EM field distribution modes for each nanostructure in the X-Y and X-Z directions was analyzed with an incident laser source with a 532 nm wavelength. As shown in Figure 3b,c, the Au NP provides a hot spot at the interface between the Au NPs and SiO_2 substrate with a maximal EM field strength ($|E|/|E_0|$) of 5.8. Figure 3e,f show the FDTD calculation of the localized EM field distribution in PMDM hybrid nanostructures with a 3 nm Al_2O_3 nanogap. The high density of the strongest hot spots is induced at the dielectric Al_2O_3 nanogap between the core Au NPs and the out-layered Au NPs due to the plasmon coupling between the Au NPs. Compared with Au NPs, the EM field strength was much higher for the PMDM hybrid nanostructures—i.e., $|E|/|E_0| \approx 21.5$. It is widely known that the SERS EF can be theoretically predicted from the local EM field enhancement ($|E|/|E_0|$) of nanostructures—i.e., SERS EF is proportional to the fourth power of $|E|/|E_0|$ [40–42]. Based on the above relation, the theoretical SERS EF is estimated to be ~2.14×10^5 for hybrid nanostructures, which is two orders higher than Au NPs. These results suggest that a high density of strong EM hotspots can be highly beneficial for SERS enhancement. It should be noted that the theoretical calculation of SERS EF from the EM enhancement is usually 2–3 orders lower than the experimental results due to the exclusion of chemical enhancement.

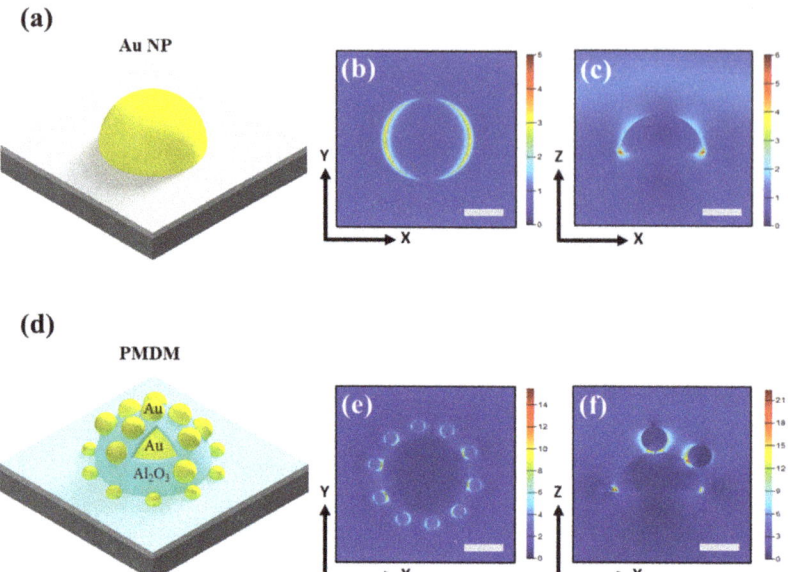

Figure 3. Finite-difference time-domain (FDTD) simulation of EM field distribution of plasmonic nanostructures under the radiation of a 532 nm laser. (**a**) Simulation model and EM field distribution of Au NP at (**b**) X-Y and (**c**) X-Z view. (**d**) Simulation model of PMDM hybrid nanostructure and corresponding EM field distribution in (**e**) X-Y and (**f**) X-Z views. Scale bar: 100 nm.

Figure 4 shows the SERS performance of PMDM hybrid nanostructures using R6G as a probe molecule and an excitation laser of wavelength 532 nm. The comparison of the SERS performance of different SERS-active substrates with a 10^{-6} M R6G concentration is demonstrated in Figure 4a, and the corresponding SERS enhancement is summarized in terms of Raman peak intensity in Figure 4b. Several of the most prominent Raman peaks of R6G are observed at the wavenumbers of 612, 776, 1185, 1310, 1363, 1506, 1574, and 1650 cm^{-1}, which are consistent with the characteristic peaks of R6G reported in the literature [43,44]. The band assignment of all the Raman peaks of R6G is also summarized in Table S1. In particular, the Raman peak intensity of the PMDM hybrid nanostructures at the wavenumber of 1650 cm^{-1} is about 3.3 and 2.2 times higher compared to that of the Au NPs and Au/Au NPs. As confirmed by the FDTD simulation results, it is obvious that the PMDM hybrid nanostructures exhibit the best performance due to the strong EM field enhancement. It is observed that the intensity of the R6G Raman signal on the hybrid nanostructure is much enhanced, as compared to that of Au NPs. Therefore, the PMDM hybrid nanostructures-based SERS substrates was further analyzed to identify the detection limit, enhancement factor, and reproducibility. Figure 4c,d show the SERS spectra of different concentrations of R6G molecules adsorbed on hybrid nanostructures in the range of 10^{-5} to 10^{-12} M. The Raman intensity is gradually reduced with the decreased R6G concentration. The lowest detectable concentration reaches 10^{-11} M, where certain Raman peaks such as 1363 and 1650 cm^{-1} can be identified, indicating that the SERS substrate possesses a high SERS sensitivity. To quantitatively study the SERS performance of hybrid nanostructures, the SERS EF was calculated using the relation EF = $(I_{SERS}/C_{SERS})/(I_R/C_R)$, where I_{SERS} and I_R correspond to the Raman peak intensities of R6G obtained from the SERS substrate and reference (SiO$_2$) substrate, whereas C_{SERS} and C_R represent the concentrations of R6G molecules on SERS substrate and reference substrate, respectively. The minimum detectable limit for the SERS substrate was 10^{-11} M, whereas the lowest detection for reference substrate was 10^{-2} M. Therefore, the SERS EF of PMDM hybrid nanostructures for Raman peak 1650 cm^{-1} was estimated as 1.3×10^8, which was much higher than that of the other SERS substrates reported in the literature (Table S2). Furthermore, we tested the reproducibility of the as-prepared SERS substrate by conducting the SERS measurement in several locations. The color contour plot of the SERS spectra of 10^{-6} M R6G measured at random 30 different locations is presented in Figure 4e and the corresponding SERS spectra are presented in Figure S2. The contour plot demonstrates the similar color of Raman signals, signifying the comparable intensity of the Raman signals due to the homogeneous distribution of EM hot spots. In addition, the SERS mapping was performed in an area of 10 μm × 10 μm to further confirm the reproducibility. The RSD values corresponding to Raman peaks 776 and 1363 cm^{-1} were calculated to be 6.8% and 7.4%, respectively, indicating the good reproducibility of SERS substrates.

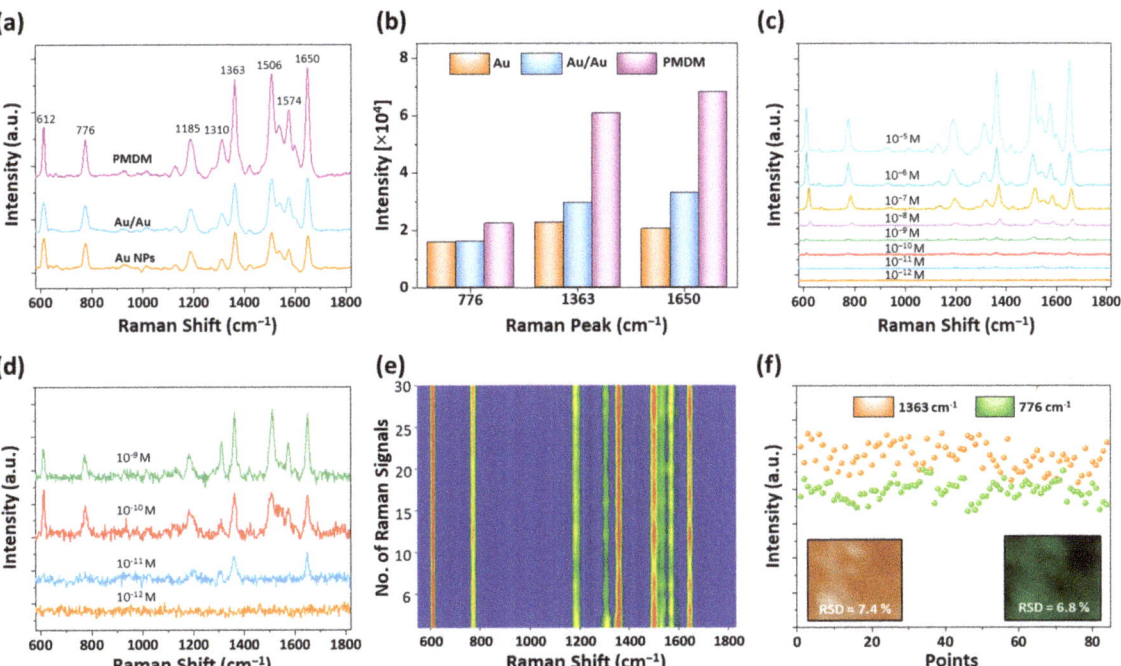

Figure 4. (a) Comparison of the SERS spectra of R6G molecules (10^{-6} M) on Au, Au/Au, and PMDM hybrid nanostructure-based SERS substrates. (b) Corresponding plot of intensity at Raman peaks of 776, 1363, and 1650 cm^{-1}. (c) SERS spectra of R6G molecules on PMDM hybrid nanostructure-based SERS substrate with different concentrations ranging from 10^{-5} to 10^{-12} M. (d) Magnified SERS spectra of R6G with low concentrations showing the distinct Raman peaks. SERS uniformity and reproducibility of the PMDM hybrid nanostructure substrate. (e) SERS contour maps of 30 spots and (f) plot of Raman intensities of 776 and 1363 cm^{-1} randomly selected from SERS mapping in an area of 10 μm × 10 μm. (Insets) SERS intensity mapping of 776 and 1363 cm^{-1}.

4. Conclusions

In summary, we developed a facile strategy for a highly sensitive and reproducible SERS substrate based on a hybrid nanostructure. A simple repeated dewetting process coupled with an ALD method was used to fabricate hierarchically assembled PMDM hybrid nano-architectures, which consist of core Au NPs and small out-layered Au NPs isolated by an Al_2O_3 layer. FDTD simulation data reveal that the use of the hybrid nanostructures leads to a high density and more intense EM hot spots through the creation of nanogaps by a dielectric spacer. Consequently, the SERS measurements of hybrids nanostructures demonstrate a maximum EF of 1.3×10^8, a low detection limit of 10^{-11} M R6G molecules, and an excellent reproducibility (RSDs less than 7.5%). Thus, we believe that hybrid SERS substrates have the potential to be used in practical applications for the highly sensitive, rapid, and reproducible detection of biomolecules.

Supplementary Materials: The following supporting information can be downloaded at: https://www.mdpi.com/article/10.3390/nano12030401/s1. Figure S1: Histogram of Au NPs arrays on Si/SiO$_2$ substrate by the annealing of 10 nm Au at 800 °C for 120 s; Figure S2: Raman spectra of R6G molecules (10−6 M) measured at 30 different locations on PMDM hybrid nanostructures to test the SERS reproducibility; Table S1: Raman band assignments of R6G molecules; Table S2: Comparison of SERS performance of current work and previously reported plasmonic NP-based SERS substrates (references [45–53] are cited in Table S2).

Author Contributions: Conceptualization, P.P. and J.I.S.; methodology, P.P., K.H.S. and M.-K.S.; software, P.P.; validation, P.P. and J.I.S.; formal analysis, P.P., M.-K.S., K.H.S. and J.I.S.; investigation, P.P., M.-K.S. and K.H.S.; resources, P.P. and J.I.S.; data curation, P.P., Y.-W.L. and J.I.S.; writing—original draft preparation, P.P., Y.-W.L. and J.I.S.; writing—review and editing, P.P., Y.-W.L. and J.I.S.; visualization, P.P.; supervision, J.I.S.; funding acquisition, P.P. and J.I.S. All authors have read and agreed to the published version of the manuscript.

Funding: This research was supported by the National Research Foundation of Korea (NRF) grant funded by the Korean government (MSIT) (2019R1A2C1007883 and 2021R1F1A1060364). This result was also supported by "Regional Innovation Strategy (RIS)" through the National Research Foundation of Korea (NRF) funded by the Ministry of Education (MOE) (2021RIS-004).

Institutional Review Board Statement: Not applicable.

Informed Consent Statement: Not applicable.

Data Availability Statement: Data are contained within the article or Supplementary Material.

Conflicts of Interest: The authors declare no conflict of interest.

References

1. Sharma, B.; Frontiera, R.R.; Henry, A.I.; Ringe, E.; Van Duyne, R.P. SERS: Materials, applications, and the future. *Mater. Today* **2012**, *15*, 16–25. [CrossRef]
2. Li, Y.; Wang, Y.; Fu, C.; Wu, Y.; Cao, H.; Shi, W.; Jung, Y.M. A simple enzyme-free SERS sensor for the rapid and sensitive detection of hydrogen peroxide in food. *Analyst* **2020**, *145*, 607–612. [CrossRef] [PubMed]
3. Plou, J.; García, I.; Charconnet, M.; Astobiza, I.; García-Astrain, C.; Matricardi, C.; Mihi, A.; Carracedo, A.; Liz-Marzán, L.M. Multiplex SERS Detection of Metabolic Alterations in Tumor Extracellular Media. *Adv. Funct. Mater.* **2020**, *30*, 1910335. [CrossRef]
4. Pandey, P.; Shin, K.; Jang, A.R.; Seo, M.K.; Hong, W.K.; Sohn, J.I. Highly sensitive multiplex-detection of surface-enhanced Raman scattering via self-assembly arrays of porous AuAg nanoparticles with built-in nanogaps. *J. Alloys Compd.* **2021**, *888*, 161504. [CrossRef]
5. Yang, H.; Gun, X.; Pang, G.; Zheng, Z.; Li, C.; Yang, C.; Wang, M.; Xu, K. Femtosecond laser patterned superhydrophobic/hydrophobic SERS sensors for rapid positioning ultratrace detection. *Opt. Express* **2021**, *29*, 16904–16913. [CrossRef]
6. Mao, P.; Liu, C.; Favraud, G.; Chen, Q.; Han, M.; Fratalocchi, A.; Zhang, S. Broadband single molecule SERS detection designed by warped optical spaces. *Nat. Commun.* **2018**, *9*, 5428. [CrossRef]
7. Blackie, E.J.; Ru, E.C.L.; Etchegoin, P.G. Single-molecule surface-enhanced Raman spectroscopy of nonresonant molecules. *J. Am. Chem. Soc.* **2009**, *131*, 14466–14472. [CrossRef]
8. Zhang, Y.; Gu, Y.; He, J.; Thackray, B.D.; Ye, J. Ultrabright gap-enhanced Raman tags for high-speed bioimaging. *Nat. Commun.* **2019**, *10*, 3905. [CrossRef]
9. Kim, J.; Nam, S.H.; Lim, D.K.; Suh, Y.D. SERS-based particle tracking and molecular imaging in live cells: Toward the monitoring of intracellular dynamics. *Nanoscale* **2019**, *11*, 21724–21727. [CrossRef]
10. Zhu, Y.; Tang, H.; Wang, H.; Li, Y. In Situ SERS Monitoring of the Plasmon-Driven Catalytic Reaction by Using Single Ag@Au Nanowires as Substrates. *Anal. Chem.* **2021**, *93*, 11736–11744. [CrossRef]
11. He, J.; Song, G.; Wang, X.; Zhou, L.; Li, J. Multifunctional magnetic Fe$_3$O$_4$/GO/Ag composite microspheres for SERS detection and catalytic degradation of methylene blue and ciprofloxacin. *J. Alloys Compd.* **2022**, *893*, 162226. [CrossRef]
12. Tian, L.; Su, M.; Yu, F.; Xu, Y.; Li, X.; Li, L.; Liu, H.; Tan, W. Liquid-state quantitative SERS analyzer on self-ordered metal liquid-like plasmonic arrays. *Nat. Commun.* **2018**, *9*, 3642. [CrossRef]
13. Lee, H.K.; Lee, Y.H.; Koh, C.S.L.; Phan-Quang, G.C.; Han, X.; Lay, C.L.; Sim, H.Y.F.; Kao, Y.-C.; An, Q.; Ling, X.Y. Designing surface-enhanced Raman scattering (SERS) platforms beyond hotspot engineering: Emerging opportunities in analyte manipulations and hybrid materials. *Chem. Soc. Rev.* **2019**, *48*, 731–756. [CrossRef]

14. Li, Z.; Jiang, S.; Huo, Y.; Ning, T.; Liu, A.; Zhang, C.; He, Y.; Wang, M.; Li, C.; Man, B. 3D silver nanoparticles with multilayer graphene oxide as a spacer for surface enhanced Raman spectroscopy analysis. *Nanoscale* **2018**, *10*, 5897–5905. [CrossRef]
15. Dai, Z.; Xiao, X.; Wu, W.; Zhang, Y.; Liao, L.; Guo, S.; Ying, J.-J.; Shan, C.-X.; Sun, M.; Jiang, C.-Z. Plasmon-driven reaction controlled by the number of graphene layers and localized surface plasmon distribution during optical excitation. *Light Sci. Appl.* **2015**, *4*, e342. [CrossRef]
16. Yin, Z.; Xu, K.; Jiang, S.; Luo, D.; Chen, R.; Xu, C.; Shum, P.; Liu, Y.J. Recent progress on two-dimensional layered materials for surface enhanced Raman spectroscopy and their applications. *Mater. Today Phys.* **2021**, *18*, 100378. [CrossRef]
17. Zhang, C.; Li, C.; Yu, J.; Jiang, S.; Xu, S.; Yang, C.; Liu, Y.J.; Gao, X.; Liu, A.; Man, B. SERS activated platform with three-dimensional hot spots and tunable nanometer gap. *Sens. Actuators B Chem.* **2018**, *258*, 163–171. [CrossRef]
18. Tian, F.; Bonnier, F.; Casey, A.; Shanahan, A.E.; Byrne, H.J. Surface enhanced Raman scattering with gold nanoparticles: Effect of particle shape. *Anal. Methods* **2014**, *6*, 9116–9123. [CrossRef]
19. Chang, Y.C.; Dvoynenko, M.M.; Ke, H.; Hsiao, H.H.; Wang, Y.L.; Wang, J.K. Double Resonance SERS Substrates: Ag Nanoparticles on Grating. *J. Phys. Chem. C* **2021**, *125*, 27267–27274. [CrossRef]
20. Markina, N.E.; Ustinov, S.N.; Zakharevich, A.M.; Markin, A.V. Copper nanoparticles for SERS-based determination of some cephalosporin antibiotics in spiked human urine. *Anal. Chim. Acta* **2020**, *1138*, 9–17. [CrossRef]
21. Quan, J.; Zhang, J.; Qi, X.; Li, J.; Wang, N.; Zhu, Y. A study on the correlation between the dewetting temperature of Ag film and SERS intensity. *Sci. Rep.* **2017**, *7*, 14771. [CrossRef]
22. Rycenga, M.; Xia, X.; Moran, C.H.; Zhou, F.; Qin, D.; Li, Z.Y.; Xia, Y. Generation of Hot Spots with Silver Nanocubes for Single-Molecule Detection by Surface-Enhanced Raman Scattering. *Angew. Chem. Int. Ed.* **2011**, *50*, 5473–5477. [CrossRef]
23. Dai, H.; Fu, P.; Li, Z.; Zhao, J.; Yu, X.; Sun, J.; Fang, H. Electricity mediated plasmonic tip engineering on single Ag nanowire for SERS. *Opt. Express* **2018**, *26*, 25031–25036. [CrossRef]
24. Li, S.; Pedano, M.L.; Chang, S.H.; Mirkin, C.A.; Schatz, G.C. Gap structure effects on surface-enhanced Raman scattering intensities for gold gapped rods. *Nano Lett.* **2010**, *10*, 1722–1727. [CrossRef]
25. Cheng, Y.W.; Wu, C.H.; Chen, W.T.; Liu, T.Y.; Jeng, R.J. Manipulated interparticle gaps of silver nanoparticles by dendron-exfoliated reduced graphene oxide nanohybrids for SERS detection. *Appl. Surf. Sci.* **2019**, *469*, 887–895. [CrossRef]
26. Wang, X.; Zhu, X.; Shi, H.; Chen, Y.; Chen, Z.; Zeng, Y.; Duan, H. Three-dimensional-stacked gold nanoparticles with sub-5 nm gaps on vertically aligned TiO_2 nanosheets for surface-enhanced Raman scattering detection down to 10 fM scale. *ACS Appl. Mater. Interfaces* **2018**, *10*, 35607–35614. [CrossRef]
27. Xu, K.; Zhang, C.; Lu, T.H.; Wang, P.; Zhou, R.; Ji, R.; Hong, M. Hybrid metal-insulator-metal structures on Si nanowires array for surface enhanced Raman scattering. *Opto-Electron. Eng.* **2017**, *44*, 185–191.
28. Hu, J.; Yu, H.; Su, G.; Song, B.; Wang, J.; Wu, Z.; Zhan, P.; Liu, F.; Wu, W.; Wang, Z. Dual-Electromagnetic Field Enhancements through Suspended Metal/Dielectric/Metal Nanostructures and Plastic Phthalates Detection in Child Urine. *Adv. Opt. Mater.* **2020**, *8*, 1901305. [CrossRef]
29. Tatmyshevskiy, M.K.; Yakubovsky, D.I.; Kapitanova, O.O.; Solovey, V.R.; Vyshnevyy, A.A.; Ermolaev, G.A.; Klishin, Y.A.; Mironov, M.S.; Voronov, A.A.; Arsenin, A.V.; et al. Hybrid Metal-Dielectric-Metal Sandwiches for SERS Applications. *Nanomaterials* **2021**, *11*, 3205. [CrossRef] [PubMed]
30. Ma, L.; Wang, J.; Huang, H.; Zhang, Z.; Li, X.; Fan, Y. Simultaneous thermal stability and ultrahigh sensitivity of heterojunction SERS substrates. *Nanomaterials* **2019**, *9*, 830. [CrossRef] [PubMed]
31. Preston, A.S.; Hughes, R.A.; Dominique, N.L.; Camden, J.P.; Neretina, S. Stabilization of Plasmonic Silver Nanostructures with Ultrathin Oxide Coatings Formed Using Atomic Layer Deposition. *J. Phys. Chem. C* **2021**, *125*, 17212–17220. [CrossRef]
32. Yang, C.; Chen, Y.; Liu, D.; Chen, C.; Wang, J.; Fan, Y.; Huang, S.; Lei, W. Nanocavity-in-multiple nanogap plasmonic coupling effects from vertical sandwich-like Au@ Al_2O_3@ Au arrays for surface-enhanced Raman scattering. *ACS Appl. Mater. Interfaces* **2018**, *10*, 8317–8323. [CrossRef]
33. Pandey, P.; Kunwar, S.; Shin, K.H.; Seo, M.K.; Yoon, J.; Hong, W.K.; Sohn, J.I. Plasmonic Core–Shell–Satellites with Abundant Electromagnetic Hotspots for Highly Sensitive and Reproducible SERS Detection. *Int. J. Mol. Sci.* **2021**, *22*, 12191. [CrossRef]
34. Dai, F.; Horrer, A.; Adam, P.M.; Fleischer, M. Accessing the hotspots of cavity plasmon modes in vertical metal–insulator–metal structures for surface enhanced Raman scattering. *Adv. Opt. Mater.* **2020**, *8*, 1901734. [CrossRef]
35. Li, M.-Y.; Yu, M.; Jiang, S.; Liu, S.; Liu, H.; Xu, H.; Su, D.; Zhang, G.; Chen, Y.; Wu, J. Controllable 3D plasmonic nanostructures for high-quantum-efficiency UV photodetectors based on 2D and 0D materials. *Mater. Horiz.* **2020**, *7*, 905–911. [CrossRef]
36. Liu, J.; Chu, L.; Yao, Z.; Mao, S.; Zhu, Y.; Lee, J.; Wang, J.; Belfiore, L.A.; Tang, J. Fabrication of Au network by low-degree solid state dewetting: Continuous plasmon resonance over visible to infrared region. *Acta Mater.* **2020**, *188*, 599–608. [CrossRef]
37. Johnson, P.B.; Christy, R.W. Optical constants of the noble metals. *Phys. Rev. B* **1972**, *6*, 4370. [CrossRef]
38. Shi, Y.; Wang, J.; Wang, C.; Zhai, T.-T.; Bao, W.-J.; Xu, J.-J.; Xia, X.-H.; Chen, H.-Y. Hot electron of Au nanorods activates the electrocatalysis of hydrogen evolution on MoS_2 Nanosheets. *J. Am. Chem. Soc.* **2015**, *137*, 7365–7370. [CrossRef]
39. Alshehri, A.H.; Mistry, K.; Nguyen, V.H.; Ibrahim, K.H.; Muñoz-Rojas, D.; Yavuz, M.; Musselman, K.P. Quantum-Tunneling Metal-Insulator-Metal Diodes Made by Rapid Atmospheric Pressure Chemical Vapor Deposition. *Adv. Funct. Mater.* **2019**, *29*, 1805533. [CrossRef]
40. Shao, F.; Lu, Z.; Liu, C.; Han, H.-Y.; Chen, K.; Li, W.; He, Q.; Peng, H.; Chen, J. Hierarchical nanogaps within bioscaffold arrays as a high-performance SERS substrate for animal virus biosensing. *ACS Appl. Mater. Interfaces* **2014**, *6*, 6281–6289. [CrossRef]

41. Xu, J.; Kvasnička, P.; Idso, M.; Jordan, R.W.; Gong, H.; Homola, J.; Yu, Q. Understanding the effects of dielectric medium, substrate, and depth on electric fields and SERS of quasi-3D plasmonic nanostructures. *Opt. Express* **2011**, *19*, 20493–20505. [CrossRef]
42. Yuan, K.; Zheng, J.; Yang, D.; Sánchez, B.J.; Liu, X.; Guo, X.; Liu, C.; Dina, N.E.; Jian, J.; Bao, Z.; et al. Self-assembly of Au@Ag nanoparticles on mussel shell to form large-scale 3D supercrystals as natural SERS substrates for the detection of pathogenic bacteria. *ACS Omega* **2018**, *3*, 2855–2864. [CrossRef]
43. Lee, T.; Jung, D.; Wi, J.S.; Lim, H.; Lee, J.J. Surfactant-free galvanic replacement for synthesis of raspberry-like silver nanostructure pattern with multiple hot-spots as sensitive and reproducible SERS substrates. *Appl. Surf. Sci.* **2020**, *505*, 144548. [CrossRef]
44. Zhao, X.; Liu, C.; Yu, J.; Li, Z.; Liu, L.; Li, C.; Xu, S.; Li, W.; Man, B.; Zhang, C. Hydrophobic multiscale cavities for high-performance and self-cleaning surface-enhanced Raman spectroscopy (SERS) sensing. *Nanophotonics* **2020**, *9*, 4761–4773. [CrossRef]
45. Zhong, F.; Wu, Z.; Guo, J.; Jia, D. Porous silicon photonic crystals coated with Ag nanoparticles as efficient substrates for detecting trace explosives using SERS. *Nanomaterials* **2018**, *8*, 872. [CrossRef]
46. Purwidyantri, A.; Hsu, C.H.; Yang, C.M.; Prabowo, B.A.; Tian, Y.C.; Lai, C.S. Plasmonic nanomaterial structuring for SERS enhancement. *RSC Adv.* **2019**, *9*, 4982–4992. [CrossRef]
47. Waiwijit, U.; Chananonnawathorn, C.; Eimchai, P.; Bora, T.; Hornyak, G.L.; Nuntawong, N. Fabrication of Au-Ag nanorod SERS substrates by co-sputtering technique and dealloying with selective chemical etching. *Appl. Surf. Sci.* **2020**, *530*, 147171. [CrossRef]
48. Xue, Y.; Paschalidou, E.M.; Rizzi, P.; Battezzati, L.; Denis, P.; Fecht, H.J. Nanoporous gold thin films synthesised via de-alloying of Au-based nanoglass for highly active SERS substrates. *Philos. Mag.* **2018**, *98*, 2769–2781. [CrossRef]
49. Zhu, J.; Wu, N.; Zhang, F.; Li, X.; Li, J.; Zhao, J. SERS detection of 4-Aminobenzenethiol based on triangular Au-AuAg hierarchical-multishell nanostructure. *Spectrochim. Acta—Part A Mol. Biomol. Spectrosc.* **2018**, *204*, 754–762. [CrossRef] [PubMed]
50. Zhou, Y.; Liang, P.; Zhang, D.; Tang, L.; Dong, Q.; Jin, S.; Ni, D.; Yu, Z.; Ye, J. A facile seed growth method to prepare stable Ag@ZrO_2 core-shell SERS substrate with high stability in extreme environments. *Spectrochim. Acta Part A Mol. Biomol. Spectrosc.* **2020**, *228*, 117676. [CrossRef]
51. Wang, J.; Li, J.; Zeng, C.; Qu, Q.; Wang, M.; Qi, W.; Su, R.; He, Z. Sandwich-like sensor for the highly specific and reproducible detection of Rhodamine 6G on a surface-enhanced Raman scattering platform. *ACS Appl. Mater. Interfaces* **2020**, *12*, 4699–4706. [CrossRef]
52. Fathima, H.; Mohandas, N.; Varghese, B.S.; Anupkumar, P.; Swathi, R.S.; Thomas, K.G. Core–Shell Plasmonic Nanostructures on Au Films as SERS Substrates: Thickness of Film and Quality Factor of Nanoparticle Matter. *J. Phys. Chem. C* **2021**, *125*, 16024–16032. [CrossRef]
53. Wang, Y.; Jin, A.; Quan, B.; Liu, Z.; Li, Y.; Xia, X.; Li, W.; Yang, H.; Gu, C.; Li, J. Large-scale Ag-nanoparticles/Al_2O_3/Au-nanograting hybrid nanostructure for surface-enhanced Raman scattering. *Microelectron. Eng.* **2017**, *172*, 1–7. [CrossRef]

Article

A Redox-Mediator-Integrated Flexible Micro-Supercapacitor with Improved Energy Storage Capability and Suppressed Self-Discharge Rate

Sung Min Wi [1,†], Jihong Kim [1,†], Suok Lee [1,†], Yu-Rim Choi [1], Sung Hoon Kim [1], Jong Bae Park [2], Younghyun Cho [1], Wook Ahn [1], A-Rang Jang [3], John Hong [4,*] and Young-Woo Lee [1,*]

[1] Department of Energy Systems Engineering, Soonchunhyang University, Asan-si 31538, Korea; dnlals77@naver.com (S.M.W.); colorg11@naver.com (J.K.); solee0117@gmail.com (S.L.); uj1008@sch.ac.kr (Y.-R.C.); nyk4123@naver.com (S.H.K.); yhcho@sch.ac.kr (Y.C.); wahn21@sch.ac.kr (W.A.)
[2] Jeonju Centre, Korea Basic Science Institute, Jeonju 54907, Korea; jbpjb@kbsi.re.kr
[3] Department of Electrical Engineering, Semyung University, Jecheon-si 27136, Korea; arjang@semyung.ac.kr
[4] School of Materials Science and Engineering, Kookmin University, Seoul 02707, Korea
* Correspondence: johnhong@kookmin.ac.kr (J.H.); ywlee@sch.ac.kr (Y.-W.L.); Tel.: +82-2-910-4665 (J.H.); +82-41-530-4988 (Y.-W.L.)
† These authors contributed equally to this work.

Abstract: To effectively improve the energy density and reduce the self-discharging rate of micro-supercapacitors, an advanced strategy is required. In this study, we developed a hydroquinone (HQ)-based polymer-gel electrolyte (HQ-gel) for micro-supercapacitors. The introduced HQ redox mediators (HQ-RMs) in the gel electrolyte composites underwent additional Faradaic redox reactions and synergistically increased the overall energy density of the micro-supercapacitors. Moreover, the HQ-RMs in the gel electrolyte weakened the self-discharging behavior by providing a strong binding attachment of charged ions on the porous graphitized carbon electrodes after the redox reactions. The micro-supercapacitors with HQ gel (HQ-MSCs) showed excellent energy storage performance, including a high energy volumetric capacitance of 255 mF cm^{-3} at a current of 1 µA, which is 2.7 times higher than the micro-supercapacitors based on bare-gel electrolyte composites without HQ-RMs (b-MSCs). The HQ-MSCs showed comparatively low self-discharging behavior with an open circuit potential drop of 37% compared to the b-MSCs with an open circuit potential drop of 60% after 2000 s. The assembled HQ-MSCs exhibited high mechanical flexibility over the applied external tensile and compressive strains. Additionally, the HQ-MSCs show the adequate circuit compatibility within series and parallel connections and the good cycling performance of capacitance retention of 95% after 3000 cycles.

Keywords: hydroquinone-based polymer-gel electrolyte; micro-supercapacitors; Faradaic redox reactions; energy storage

1. Introduction

Recent studies have demonstrated the potential of flexible micro-supercapacitors for supplying energy and electricity to future flexible and wearable electronics such as rollable displays, human-implanted devices, and high-end robotics [1–3]. The micro-supercapacitors are highly significant as future energy storage devices because they can be integrated with small-sized applications, operate under fast charge/discharge conditions, and have a long lifetime [4]. Moreover, developing an effective method to fabricate electrode structures on flexible substrates and depositing electrode materials on small areas is crucial for the successful utilization of micro-supercapacitors. As a result, tremendous efforts have been directed to develop carbon-based micro-supercapacitor electrode materials [5].

Carbon materials can be easily handled on flexible substrates, and their electrical and chemical properties are well tailored by a simple post-treatment process, inducing high electrochemical energy storage performance [6,7]. For example, the gold and nitrogen doping on the carbon electrode sample can increase the conductivity and wettability of the carbon electrode, inducing the improved electrochemical performance [8]. Moreover, Peng et al. reported that the boron doped laser-induced graphene has highly improved electrochemical performance, greater than the pure laser-induced graphene [9]. However, carbon-based micro-supercapacitors inevitably have a lower energy density than other energy storage systems because of their electrostatic/physical-only charge-storing kinetics [10,11]. In general, the energy density of carbon-based electrodes based on electric double layer capacitor (EDLC) lies in the range of 0.1~3 Wh kg^{-1} [12], but in a range of over 100 Wh kg^{-1} for Li ion batteries. There is also another type of supercapacitor (pseudocapacitors, with an energy density of about 10 Wh kg^{-1}), but they store charges through Faradaic redox reactions on the surface of electrodes [13]. Moreover, carbon materials for flexible micro-supercapacitors based on EDLC suffer from a high self-discharging rate owing to the weak attachment of electrolyte ions on the carbon electrodes. Additionally, the polymer-gel electrolyte is another essential component to develop flexible and wearable micro-supercapacitors [14,15]. In general, the classic liquid-type electrolyte has critical issues to apply the flexible and wearable supercapacitors due not only to their electrolyte leakage but also to their high manufacturing costs, such as difficult packaging to fabricate flexible supercapacitors [16,17]. However, the pure polymer-gel electrolyte has the low ionic conductivity of the polymer medium [18,19]. Therefore, enhancing the energy storage performance and minimizing the self-discharging behavior are critical issues that must be resolved for carbon-based flexible micro-supercapacitors.

The prevalent carbon materials used in micro-supercapacitors are graphite-based 2D planar materials because of their outstanding electrical conductivity, highly tunable surface area, chemical stability, and mechanical behavior [20,21]. Therefore, many scientists have studied tailoring the three-dimensional morphology and surface functionalization of graphite materials to enhance their electrochemical properties [22,23]. Another strategy being investigated is the use of redox mediators (RMs) in gel electrolytes [24–26]. Especially, RMs can show high flexibility and mechanical/chemical stability when they are mixed with a gel electrolyte, as well as provide easy diffusion in the gel electrolyte. The addition of RMs plays pivotal roles in enhancing the performance of supercapacitors due to the induced electrochemical Faradaic redox reactions on the surface of electrodes, which can store more electron charges compared to double-layer capacitance [27,28]. Thus, the total capacitance of supercapacitors with redox mediators can store energy by both electric double layer capacitance and the pseudocapacitance working in parallel. Additionally, interestingly, RMs play key roles in minimizing the self-discharging behavior. In particular, Faradaic redox reactions of RMs result in a high binding attachment level of charged ions on carbon-based electrodes, and RMs increase the ionic conductivity of the gel electrolyte, inducing a low self-discharging rate. Therefore, introducing gel electrolyte composites with proper redox mediators might be crucial to further maximize the performance of carbon-based micro-supercapacitors. Especially, among various RMs, hydroquinone compounds can be regarded as one of the most promising redox-active mediators due to its small size and high electrochemical reversibility.

In this study, inspired by the highly interactive hydroquinone redox mediators (HQ-RMs), we systematically engineered composite mixtures with hydroquinone (HQ) as a redox mediator, polyvinyl alcohol (PVA) as a polymer-gel medium, and phosphoric acid as an acidic electrolyte (HQ-gel). The interdigitated graphite electrodes were fabricated by carbonization of polyimide (PI) sheets using a laser scribing method. The laser scribing method can be operated with a simple step process on polymer films (fast processing time) with good reproducibility by the systematic control of laser beams. Additionally, continuous fabrication on the polymer sheets is available. Finally, the carbon electrode materials can be simply deposited on the interdigitated structure by the induced carbonization from the polymer films. The assembled micro-supercapacitors with HQ-gel (HQ-MSCs) exhibit superior electrochemical performance, including a high volumetric capacitance of 255 mF cm^{-3}, low self-discharge rate of an open circuit potential drop of 37% after 2000 s, and over 95% capacitance retention over 3000 charge/discharge cycles compared to the MSCs without the HQ-RMs (a volumetric capacitance of 94 mF cm^{-3}, self-discharging rate of an open circuit potential drop of 50% after 2000 s, and 90% capacitance retention over 3000 charge/discharge cycles). This enhancement might be attributed to the Faradaic redox reactions by the HQ-RMs and the strengthened adsorption of charged electrolyte ions on the carbon-based electrode. These findings demonstrate that the novel HQ-based gel electrolyte composites can be used to guarantee flexible carbon-based micro-supercapacitors with promising electrochemical energy storage performance for future wearable energy storage applications.

2. Materials and Methods

2.1. Fabrications of HQ-MSCs

For the HQ-MSCs, interdigitated carbon-based electrodes were directly fabricated by carbonization on PI films using a laser scribing method. The interdigitated carbon-based electrodes have seven fingers, and each electrode serves as both a working electrode and a current collector. This system does not require any separator because the interdigitated carbon-based electrodes are already separated on the PI film substrate with a length of 0.5 mm. For the electrolyte coating method, we prepared HQ-based polymer–gel electrolyte composites consisting of HQ (0.6 g, Sigma-Aldrich, Saint Louis, MO, USA) as a redox mediator, poly(vinyl alcohol) (PVA, Mw: 89,000–98,000, Sigma-Aldrich, Saint Louis, MO, USA), phosphoric acid (H_3PO_4, Sigma-Aldrich, Saint Louis, MO, USA), and deionized water (20 mL). The prepared HQ-based polymer–gel electrolyte was coated onto the interdigitated carbon-based electrodes and then dried overnight for stabilization.

2.2. Characterization and Electrochemical Tests of HQ-MSCs

We carried out powder XRD (Miniflex 600, Rigaku), Raman spectroscopy (iXR raman in Nexsa XPS system, Thermo Scientific, Korea Basic Science Institute-Jeonju Center), XPS (Nexsa XPS system, Thermo Scientific, Korea Basic Science Institute-Jeonju Center), and field-emission scanning electron microscopy (FE-SEM, Gemini SEM 300, ZEISS, Jena, Germany) analyses. In addition, the BET surface area of the samples was measured using nitrogen adsorption/desorption measurements (Belsorp mini X, MicrotracBEL Corp., Osaka, Japan). To confirm the deposition of the HQ-RMs, we performed Fourier transform infrared spectroscopy (FT-IR, TENSOR27, Bruker, NCIRF, Seoul National University-National Center for Inter-University Research Facilities, Billerica, MA, USA) analysis. The electrochemical capacitive behavior of the as-prepared MSCs was estimated using a potentiostat (PGSTAT302N, Metrohm, Autolab). The specific capacitance of the carbon electrodes was calculated by the GCD discharge curves. The specific areal capacitance was calculated by the discharge time and current density (mA/unit area), and the calculated specific areal capacitance was divided by the electrode thickness to evaluate the specific volumetric capacitance of the samples.

3. Results and Discussion

As shown in Figure 1a, the interdigitated carbon-based electrodes were fabricated using a stepwise direct laser scribing method on polyimide (PI) sheets. With direct laser irradiation, carbonization of the PI sheets immediately occurs using a pulsed laser and forms carbon-based electrodes. The interdigitated carbon-based electrodes were scribed on the PI sheets. After the laser-carbonization process, the HQ-gel composites were drop-coated onto the interdigitated carbon-based electrodes. Finally, the interdigitated carbon-based micro-supercapacitors with HQ-gel (HQ-MSCs) were dried overnight to stabilize the gel electrolyte. The interdigitated electrode structure used in the micro-supercapacitors is shown schematically in Figure 1b,c. Each finger has been designed by the fixed two-dimensional interdigitated structure (length of 7.5 mm and finger width of 1 mm). The gap distance between neighboring finger electrodes is ~0.5 mm. According to the cross-sectional SEM images (Figure S1), the electrodes show a thickness of 12 μm and the electrolyte layers have a thickness of 14 μm. Figure 1d shows the optical images of the fabricated HQ-MSCs on the PI sheets. Owing to the high flexibility of the PI sheets, the HQ-MSCs can sustain their original interdigitated MSC structure even when the PI substrates are strongly subjected to external bending forces. A cross-sectional schematic of the HQ-MSCs is shown in Figure 1e. On the PI substrates, two unconnected carbon-based electrodes were assigned to the symmetric anode and cathode electrodes. Both electrodes were covered by the HQ-gel composites. During the charge and discharge processes, the existence of the HQ-RMs in the polymer–gel electrolyte induces additional Faradaic redox reactions and delivers a high energy density compared to the micro-supercapacitors with the bare-gel electrolyte.

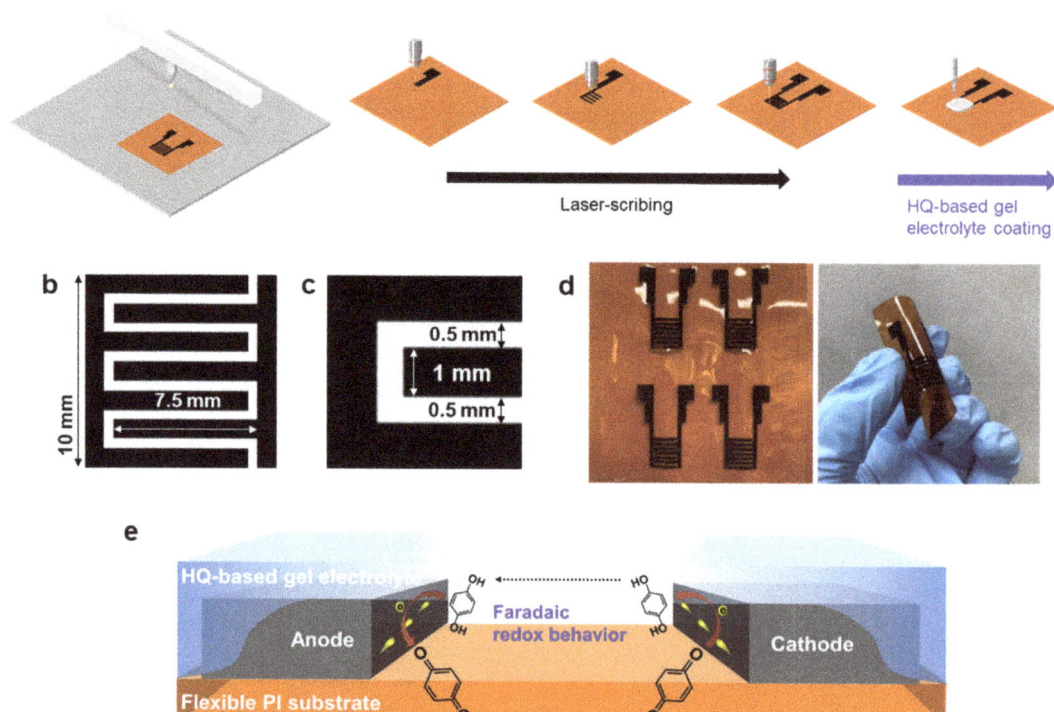

Figure 1. (a) Illustration of fabrication process for the interdigitated carbon-based micro-supercapacitors with HQ-gel (HQ-MSCs) based on a laser scribing method and HQ-based gel electrolyte. (b,c) Schematic image for pattern structure of MSCs. (d) Photograph of MSCs prepared by a laser scribing process. (e) Schematic illustration of electrochemical Faradaic redox behaviors of HQ in HQ-MSCs.

To evaluate the carbonization process by the crystallographic phase, X-ray diffraction (XRD) spectra of the pure PI films and carbon-based electrodes from the PI films were analyzed (Figure 2a). The clear XRD peaks at 15° and 27°, as well as broad intensity areas near 22.5°, were well matched with the crystal phase of PI [29]. After laser irradiation, new peaks at 23° were ascribed to the graphite-like carbon crystals after the carbonization process. The XRD spectrum of the graphite-like carbon-based electrodes (GCEs) exhibited a slight negative shift compared to that of the intrinsic graphite index. The peak shift can be attributed to the expanded d-spacing value of the GCEs resulting from the partial formation of the oxygen-containing functional group on the graphite layers during the laser carbonization process [30,31]. Figure 2b shows the Raman spectrum of the GCEs with three strong peaks at 1346 cm^{-1} (D-band), 1584 cm^{-1} (G-band), and 2689 cm^{-1} (2D-band) [32,33]. In general, the high ratio of I_{2D}/I_G indicates the typical features of graphene. As the number of graphene layers increases, the ratio of I_{2D}/I_G decreases. Thus, the I_{2D}/I_G of graphite is commonly lower than that of graphene, which is less than 1 [34–36]. In this work, the I_{2D}/I_G of GCEs is approximately 0.746, which demonstrates that the PI films were converted to graphite composites with two-dimensional layered structures. The existence of strong 2D-band peaks demonstrates that the PI films were converted to graphite composites with two-dimensional layered structures.

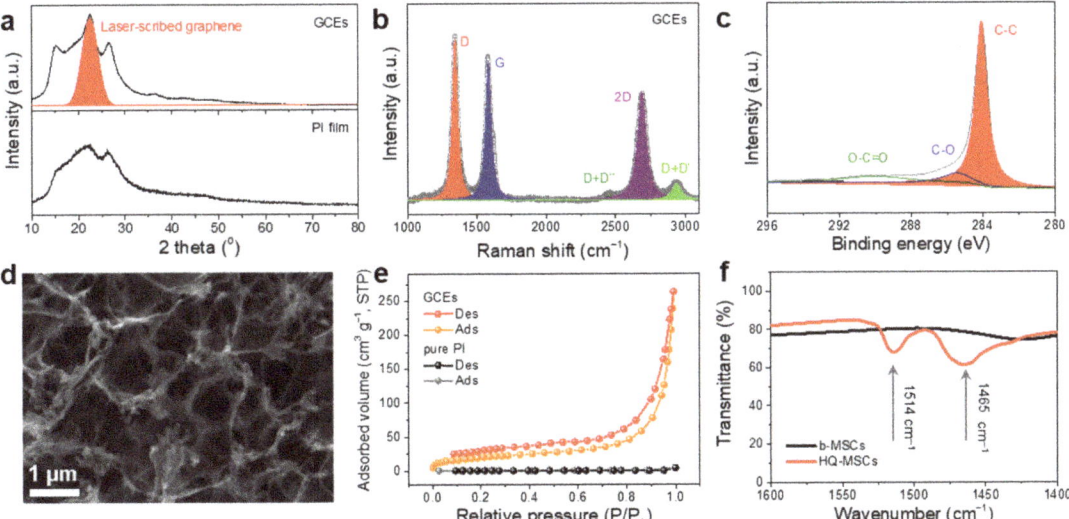

Figure 2. (**a**) X-ray diffraction (XRD) patterns of graphite-like carbon-based electrodes (GCEs) and pure polyimide (PI) film. (**b**) Raman spectrum and (**c**) X-rya photoelectron spectroscopy (XPS) C1s spectrum of GCEs. (**d**) Scanning electron microscopy (SEM) image of GCEs. (**e**) Nitrogen adsorption/desorption isotherm curves of GCEs and pure PI film. (**f**) Fourier transform infrared spectroscopy (FT-IR) spectra of HQ-MSCs and b-MSCs.

We also carried out X-ray photoelectron spectroscopy (XPS) measurements to determine the surface properties of the GCEs (Figure 2c). The GCEs exhibited a common graphite characteristic with a strong C-C peak clearly observed at 284.6 eV, which indicates a high degree of formation of layered graphite structures [37–39]. In addition, as shown in Figure 2d, the as-prepared GCEs showed a highly porous structure based on a three-dimensional network, owing to the rapid formation of gaseous species produced during laser irradiation. To further evaluate the specific surface area of the GCEs, Brunauer–Emmett–Teller (BET) measurements were carried out. As shown in Figure 2e, the GCEs exhibited typical adsorption/desorption curves of type II [40], and the calculated BET surface areas of the GCEs were observed to be 74.15 m^2 g^{-1}, which is higher than that of pure PI (0.08 m^2 g^{-1}). After the laser patterning process, the PI substrates were successfully converted to layered GCEs with large surface sites and good electrical conductivity, which are favorable for electrochemical energy storage. The measured electrical resistance of the GCEs by using the 2-probe method is approximately 55.6 Ω, whereas the resistance of the pure PI films was not measured due to the insulating properties (Figure S2).

In addition, the existence of the HQ-RMs in the polymer-gel electrolyte was investigated by Fourier transform infrared spectroscopy (FT-IR) spectra (Figure 2f). The HQ-MSCs showed two dominant peaks in the FT-IR spectrum, which corresponded to the phenyl ring stretching (1512 cm^{-1}) and -C-OH in-plane bending (1465 cm^{-1}) of HQ compared to MSCs without HQ-RMs (b-MSCs) [41]. These two clear peaks are the main signs that the HQ-RMs are well mixed in the gel electrolyte composites and are deposited on the GCEs. Thus, it is expected that the HQ-MSCs will exhibit improved energy-storing properties owing to their unique structural/electrochemical features as follows: (1) the well-designed graphite-like carbon-based interdigitated electrodes with good electrical conductivity, which supports the fast electron pathway; (2) the large surface area by porous structures, which provide large electrolyte contact areas; and (3) the induction of additional Faradaic redox reactions using HQ-RMs, which induce improved energy storage properties, as shown in Figure 1e. There is synergistic electrochemical contribution on the surface of the carbon electrodes (both electrical double-layer capacitance and Faradaic redox reactions by the HQ-RMs) [42–44].

The electrochemical properties of the HQ-MSCs were evaluated using a two-electrode system. Figure 3a shows the cyclic voltammetry (CV) curves of the HQ-MSCs and b-MSCs at a scan rate of 100 mV s^{-1}. The area surrounded by the CV curve of the HQ-MSCs was larger than that of the B-MSCs, demonstrating a higher energy storage performance of the HQ-RMs. The CV of the HQ-MSCs exhibited a pair of peaks at 0.15 V, which is a significant characteristic of the electrochemical multiple Faradaic redox reaction of the HQ-RMs during the charge/discharge cycles. The expected redox reactions of the HQ-RMs during the charge/discharge cycles are as follows (Figure 3b) [23,41,43,45]:

hydroquinone (HQ) → benzoquinone (BQ) 2H$^+$ + 2e$^-$ (charge process)
benzoquinone (BQ) 2H$^+$ + 2e$^-$ → hydroquinone (HQ) (discharge process)

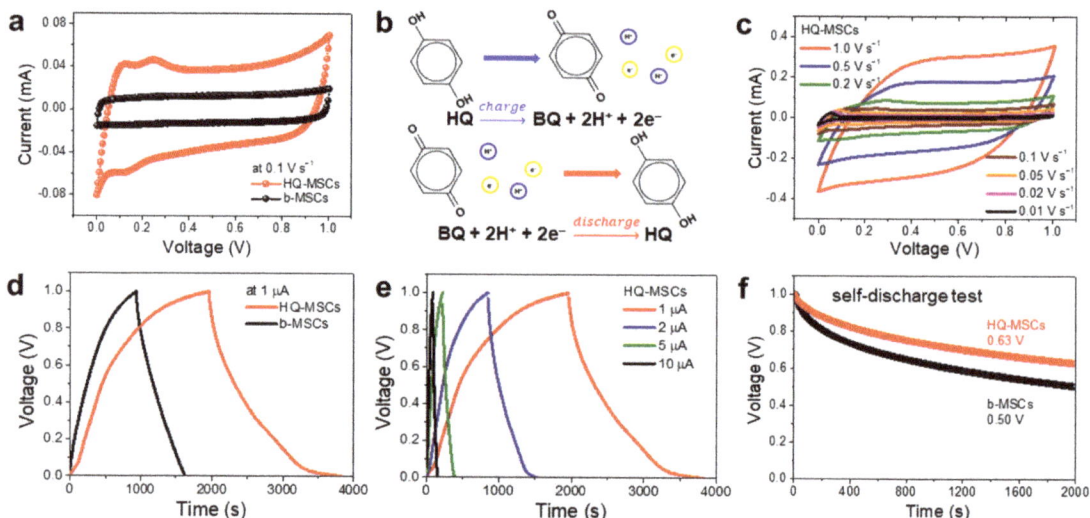

Figure 3. (**a**) Cyclic voltammetry (CV) curves of HQ-MSCs and b-MSCs at a scan rate of 0.1 V s^{-1}. (**b**) Electrochemical Faradaic redox behavior mechanism between HQ and BQ. (**c**) CV curves of HQ-MSCs at different scan rates from 0.01 to 1.0 V s^{-1}. (**d**) Galvanic charge/discharge graph (GCD) curves of HQ-MSCs and b-MSCs at a current 1 µA. (**e**) GCD curves of HQ-MSCs at different current ranges from 1 to 10 µA. (**f**) Self-discharging test of HQ-MSCs and b-MSCs after fully charged state.

Furthermore, the CV curves of the HQ-MSCs showed similar shapes with increasing scan rates from 10 to 100 mV s^{-1}, indicating that the HQ-MSCs have good energy-storing kinetics and reversible capacitive behavior (Figure 3c). Especially, as indicated in Figure S3, at slow scan rates, all the possible ion adsorption and electrochemical reactions are maximized on the surface within the given sweeping window (the clear redox pairs are detected). However, at the fast scan rates, the relatively broad redox peaks can be observed as shown in Figures 3c and S3, which are the normally recognized CV response of Faradaic redox materials. Furthermore, we carried out the CV tests to confirm the effects of the concentration of HQ in the MSCs. As shown in Figure S4, the HQ-MSCs with a HQ concentration of 0.135 M showed the low energy storing performance, which is 8.1 times lower than that of HQ-MSCs with the HQ concentration of 0.27 M. Furthermore, the enclosed CV areas of HQ-MSCs with an HQ concentration of 0.54 M (the maximum aqueous solubility) is also smaller than that of HQ-MSCs with an HQ concentration of 0.27 M. The excess amount of HQ-RMs in the electrolyte can decrease the overall electrochemical performance due to the low ionic conductivity and ion permeability through the gel electrolyte [44].

The galvanic charge/discharge graph (GCD) of the HQ-MSCs presented a longer discharge time than that of the b-MSCs (Figure 3d,e). The calculated area capacitance of the HQ-MSCs (2.58 mF cm^{-2}) was approximately 2.7 times higher than that of the B-MSCs (0.95 mF cm^{-2}) and other previously reported studies (summarized in Figure S5a). In addition, the volumetric capacitance of HQ-MSCs is 255 mF cm^{-3}. The improved energy storage properties of HQ-MSCs were attributed to the additional Faradaic redox reactions of the HQ-RMs compared to those of the bare-gel electrolyte composites. In addition, as mentioned in the introduction, self-discharge in carbon-based micro-supercapacitors is another important issue that must be addressed to develop high-performance MSCs. To compare the self-discharging rate between HQ-MSCs and b-MSCs, we measured the voltage drop based on the rest time from the fully charged state of MSCs. As shown in Figure 3f, the HQ-MSCs exhibited a low self-discharge rate. The open circuit voltage drop rate of the HQ-MSCs was 37% after 2000s, which was lower than that of b-MSCs (50%) and other previously reported studies (summarized in Figure S5b). The charged ions formed by the electrochemical Faradaic redox reaction of the HQ-RMs were strongly adsorbed on the electrodes and had a low free diffusion rate into the bulk electrolyte under the polymer-gel electrolyte; therefore, the HQ-MSCs can exhibit low self-discharge rate behavior. Previous studies reported that a polymer–gel electrolyte with limited moisture exhibited a superiorly suppressed self-discharge rate compared to aqueous electrolytes because the limited moisture condition decreased the level of ion mobility from the surface of the electrode to the bulk electrolyte solution [46–50]. In addition, the charged species of the HQ-RMs formed during electrochemical capacitive behavior were adsorbed on the electrode surface, thereby suppressing the self-discharge process [49]. When the self-discharge rate of the MCSs was tested under the aqueous electrolyte solution with HQ, the open circuit voltage drop rate of the HQ-MSCs exhibited a rapid voltage drop within 200 s, as shown in Figure S6.

Furthermore, the mechanical flexibility and stability of the HQ-MSCs were estimated under different external strain levels (measured radius of curvature). The levels of external strains were normalized by the radius of curvature from 5 to 10 mm (Figure 4a,b). Figure 4c presents the CV curves of the HQ-MSCs when different levels of external strains were applied at a scan rate of 500 mV s^{-1}. The CV curves did not show any significant changes during the strain tests, indicating its superior mechanical flexibility and stability against external strains. Furthermore, the HQ-MSCs exhibited a superior mechanical stability with a capacitance retention of 99.1% during the 1000 bending cycles (Figure S7). In addition, the mechanical stability of the HQ-MSCs against the tensile and compressive strains (at a radius of curvature of 5 mm) was analyzed (Figure 4d). Optical images of the HQ-MSCs under tensile and compressive strains are shown in Figure 4e. The CV curves at a scan rate of 500 mV s^{-1} under the tensile and compressive strains also have a similar shape to the CV curves without any significant curve distortion. After the diverse external strain tests, it can be clearly confirmed that the HQ-MSCs can be successfully applied to wearable devices owing to their high flexibility and performance stability against external forces.

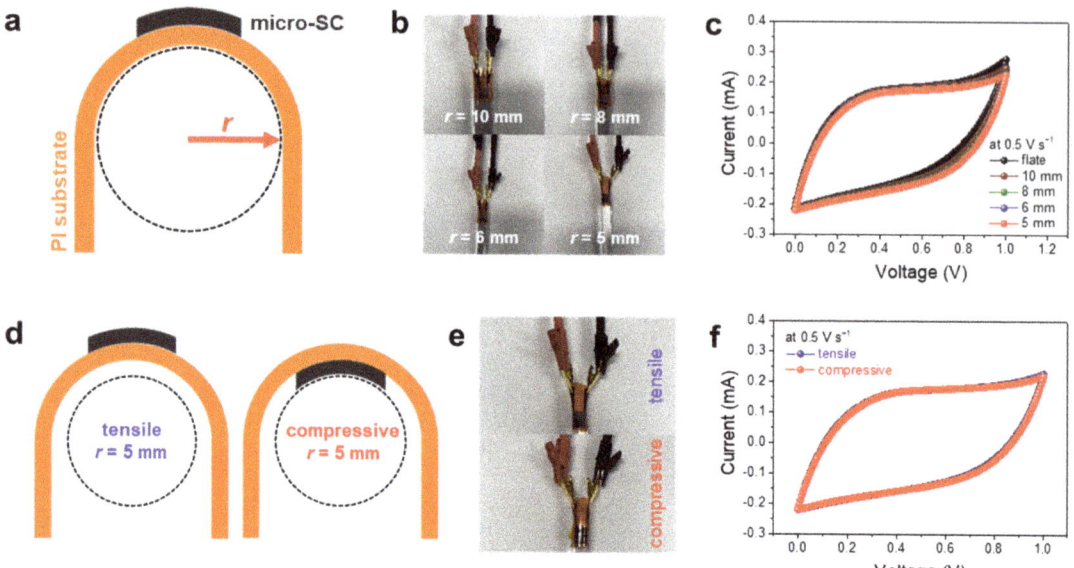

Figure 4. (**a**) Schematic illustration of bending states for HQ-MSCs. (**b**) Photographs and (**c**) CV curves of various bending states with bending radii of 5, 6, 8, and 10 mm for HQ-MSCs. (**d**) Schematic illustration and (**e**) photographs of HQ-MSCs under tensile and compressive strains. (**f**) CV curves of HQ-MSCs under tensile and compressive strains.

To evaluate the circuit applicability of the HQ-MSCs, four different HQ-MSCs were assembled in series and parallel (Figure 5a). With the series or parallel connection, the operating cell voltage or capacitance is expected to increase proportionally according to the number of connected HQ-MSCs. Figure 5b shows the CV curves when the HQ-MSCs were connected in a series circuit. The voltage windows of 1 cell to 4 cells in series increased from 1 to 4 V, respectively. In addition, the CV currents increased proportionally based on the currents from 0.17 to 0.63 mA at 0.4 V when the HQ-MSCs were connected in parallel (Figure 5c). These CV results in the series and parallel circuits demonstrated great circuit operation of the HQ-MSCs when they were utilized as potential future energy-storing devices. Moreover, as shown in Figure 5d, HQ-MSCs show great reproducibility in energy-storing performance owing to their programmed fabrication process based on the laser scribing method of MSCs. The HQ-MSCs exhibited a promising capacitive retention behavior of 95% after 3000 cycles (Figure 5e).

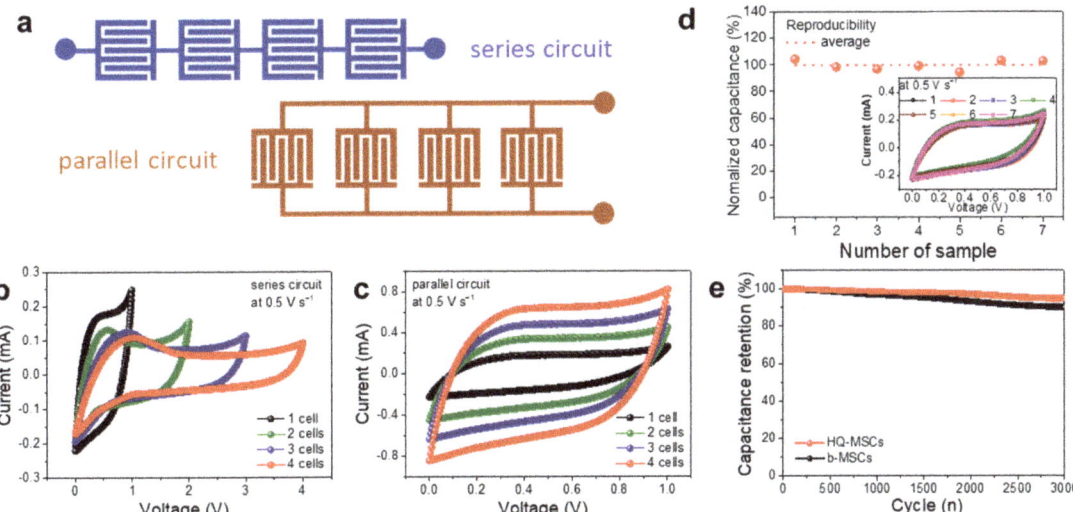

Figure 5. (**a**) Schematic illustration of series and parallel circuits for HQ-MSCs. CV curves of HQ-MSCs connected in (**b**) series and (**c**) parallel circuits. (**d**) Reproducibility and (**e**) cyclability tests of HQ-MSCs (the inset of (**d**) indicates CV curves obtained for seven HQ-MSCs devices).

4. Conclusions

In this study, we fabricated carbon-based micro-supercapacitors with an HQ-RM gel electrolyte using the laser scribing method for electrode patterning and drop-coating for MSC assembly. The carbon-based interdigitated electrodes formed using the laser scribing method show graphitic carbon crystalline features with high electrical conductivity and a porous structure. In terms of electrochemical features, HQ-MSCs have a high volumetric capacitance of 255 mF cm^{-3} at a current of 1 µA, which is 2.7 times higher than that of the b-MSCs, as well as a low self-discharge rate with an open circuit potential drop of 37% after 2000 s. The corresponding results are highly relevant for performing additional Faradaic redox reactions of the HQ-RMs and synergistically improve the overall energy storage performance. Moreover, the HQ-RMs in the gel electrolyte decrease the self-charging rate by providing a strong binding attachment of electrolyte ions on the surface of the electrodes. Furthermore, the HQ-MSCs displayed remarkably excellent mechanical features under various external mechanical stresses. Additionally, the HQ-MSCs exhibited a high reproducibility and a long-term cyclability with a high cycling capacitance retention of 95% after 3000 cycles. Therefore, the introduction of HQ redox mediators in micro-supercapacitor systems is a promising gel electrolyte additive for flexible high-performance energy storage applications.

Supplementary Materials: The following are available online at https://www.mdpi.com/article/10.3390/nano11113027/s1, Figure S1: Cross-section SEM images of the fabricated MSCs, Figure S2: Photograph images of the measured two-probe electrical resistance: (left) the GCEs region and (right) the pure PI film for the patterned MSCs, Figure S3: CV curves of HQ-MSCs with different scan rates from 0.1 V s^{-1} and to 1.0 V s^{-1}, Figure S4: CV curves of HQ-MSCs with different HQ concentrations of (a) 0.27 M and (b) 0.54 M at a scan rate of 1.0 V s^{-1}, respectively, Figure S5: Comparison of the areal capacitance and self-discharge rate of the fabricated MSCs compared to other previously reported literatures, Figure S6: Comparison of self-discharging test of MSCs under gel electrolyte and aqueous electrolyte with HQ after fully charged state, Figure S7. Capacitance retention of HQ-MSCs for 1000 bending cycles (inset indicates the CV curves of HQ-MSCs).

Author Contributions: Conceptualization, formal analysis, data curation, S.M.W., J.K. and S.L.; Formal analysis, Y.-R.C. and S.H.K.; Data curation, J.B.P.; Investigation, Y.C., W.A. and A.-R.J.; Supervision, Writing—original draft, Writing—review and editing, J.H. and Y.-W.L. All authors have read and agreed to the published version of the manuscript.

Funding: This research was supported by the Soonchunhyang University Research Fund and the Korea Institute of Energy Technology Evaluation and Planning (KETEP) and the Ministry of Trade, Industry, and Energy (MOTIE) of the Republic of Korea (No. 20184030202130).

Conflicts of Interest: The authors declare no conflict of interest.

References

1. Li, F.; Li, Y.; Qu, J.; Wang, J.; Bandari, V.K.; Zhu, F.; Schmidt, O.G. Recent developments of stamped planar micro-supercapacitors: Materials, fabrication and perspectives. *Nano Mater. Sci.* **2021**, *3*, 154–169. [CrossRef]
2. Bu, F.; Zhou, W.; Xu, Y.; Du, Y.; Guan, C.; Huang, W. Recent developments of advanced micro-supercapacitors: Design, fabrication and applications. *NPJ Flex. Electron.* **2020**, *4*, 31. [CrossRef]
3. Zhang, H.; Cao, Y.; Chee, M.O.L.; Dong, P.; Ye, M.; Shen, J. Recent advances in micro-supercapacitors. *Nanoscale* **2019**, *11*, 5807–5821. [CrossRef]
4. Maphiri, V.M.; Rutavi, G.; Sylla, N.F.; Adewinbi, S.A.; Fasakin, O.; Manyala, N. Novel thermally reduced graphene oxide microsupercapacitor fabricated via mask—Free AxiDraw direct writing. *Nanomaterials* **2021**, *11*, 1909. [CrossRef] [PubMed]
5. Liu, L.; Ye, D.; Yu, Y.; Liu, L.; Wu, Y. Carbon-based flexible micro-supercapacitor fabrication via mask-free ambient micro-plasma-jet etching. *Carbon* **2017**, *111*, 121–127. [CrossRef]
6. Smith, A.D.; Li, Q.; Vyas, A.; Haque, M.M.; Wang, K.; Velasco, A.; Zhang, X.; Thurakkal, S.; Quellmalz, A.; Niklaus, F.; et al. Carbon-based electrode materials for microsupercapacitors in self-powering sensor networks: Present and future development. *Sensors* **2019**, *19*, 4231. [CrossRef] [PubMed]
7. Fan, J.-Q.; Tu, Q.-M.; Geng, C.-L.; Huang, J.-L.; Gu, Y.; Lin, J.-M.; Huang, Y.-F.; Wu, J.-H. High energy density and low self-discharge of a quasi-solid-state supercapacitor with carbon nanotubes incorporated redox-active ionic liquid-based gel polymer electrolyte. *Electrochim. Acta* **2020**, *331*, 135425. [CrossRef]
8. Singh, B.K.; Shaikh, A.; Dusane, R.O.; Parida, S. Nanoporous gold–Nitrogen–doped carbon nano-onions all-solid-state microsupercapacitor. *Nano Struct. Nano Objects* **2019**, *17*, 239–247. [CrossRef]
9. Peng, Z.; Ye, R.; Mann, J.A.; Zakhidov, D.; Li, Y.; Smalley, P.R.; Lin, J.; Tour, J.M. Flexible boron-doped laser-induced graphene microsupercapacitors. *ACS Nano* **2015**, *9*, 5868–5875. [CrossRef]
10. Poonam; Sharma, K.; Arora, A.; Tripthi, S.K. Review of supercapacitors: Materials and devices. *J. Energy Storage* **2019**, *21*, 801–825. [CrossRef]
11. Najib, S.; Erdem, E. Current progress achieved in novel materials for supercapacitor electrodes: Mini review. *Nanoscale Adv.* **2019**, *1*, 2817–2827. [CrossRef]
12. Rani, J.R.; Thangavel, R.; Oh, S.-I.; Lee, Y.S.; Jang, J.-H. An ultra-high-energy density supercapacitor; fabrication based on thiol-functionalized graphene oxide scrolls. *Nanomaterials* **2019**, *9*, 148. [CrossRef] [PubMed]
13. Jiang, Y.; Liu, J. Definitions of pseudocapacitive materials: A brief review. *Energy Envion. Mater.* **2019**, *2*, 30–37. [CrossRef]
14. Alipoori, S.; Mazinani, S.; Aboutalebi, S.H.; Sharif, F. Review of PVA-based gel polymer electrolytes in flexible solid-state supercapacitors: Opportunities and challenges. *J. Energy Storage* **2020**, *27*, 101072. [CrossRef]
15. Jin, M.; Zhang, Y.; Yang, C.; Fu, Y.; Guo, Y.; Ma, X. High-performance ionic liquid-based gel polymer electrolyte incorporating anion-trapping boron sites for all-solid-state supercapacitor application. *ACS Appl. Mater. Interfaces* **2018**, *10*, 39570–39580. [CrossRef] [PubMed]
16. Seol, M.-L.; Nam, I.; Sdatian, E.; Dutta, N.; Han, J.-W.; Meyyappan, M. Printable gel polymer electrolytes for solid-state printed supercapacitors. *Nanomaterials* **2021**, *14*, 316. [CrossRef] [PubMed]
17. Liu, L.; Dou, Q.; Sun, Y.; Lu, Y.; Lu, Y.; Zhang, Q.; Meng, J.; Zhang, X.; Shi, S.; Yan, X. A moisture absorbing gel electrolyte enables aqueous and flexible supercapacitors operating at high temperatures. *J. Mater. Chem. A* **2019**, *7*, 20398–20404. [CrossRef]
18. Chen, L.; Bai, H.; Huang, Z.; Li, L. Mechanism investigation and suppression of self-discharge in active electrolyte enhanced supercapacitors. *Energy Environ. Sci.* **2014**, *7*, 1750–1759. [CrossRef]
19. Xia, M.; Nie, J.; Zhang, Z.; Lu, X.; Wang, Z.L. Suppressing self-discharge of supercapacitors via electrorheological effect of liquid crystals. *Nano Energy* **2018**, *47*, 43–50. [CrossRef]
20. Lin, J.; Peng, Z.; Liu, Y.; Ruiz-Zepeda, F.; Ye, R.; Samuel, E.L.G.; Yacaman, M.J.; Yakobson, B.I.; Tour, J.M. Laser-induced porous graphene films from commercial polymers. *Nat. Commmun.* **2014**, *5*, 5714. [CrossRef]
21. Tran, T.S.; Dutta, N.K.; Choudhury, N.R. Graphene-based inks for printing of planar micro-supercapacitors: A review. *Materials* **2019**, *12*, 978. [CrossRef] [PubMed]
22. Kim, C.; Kang, D.-Y.; Moon, J.H. Full lithographic fabrication of boron-doped 3D porous carbon patterns for high volumetric energy density microsupercapacitors. *Nano Energy* **2018**, *53*, 182–188. [CrossRef]
23. Chen, Y.-C.; Lin, L.-Y. Investigating the redox behavior of activated carbon supercapacitors with hydroquinone and p-phenylenediamine dual redox additives in the electrolyte. *J. Colloid Interface Sci.* **2019**, *537*, 295–305. [CrossRef] [PubMed]

24. Nagar, B.; Dubal, D.P.; Pies, L.; Merkoçi, A.; Gómez-Romero, P. Design and fabrication of printed paper-based hybrid micro-supercapacitor by using graphene and redox-active electrolyte. *ChemSusChem* **2018**, *11*, 1849–1856. [CrossRef]
25. Akinwolemiwa, B.; Peng, C.; Chen, G.Z. Redox electrolytes in supercapacitors. *J. Electrochem. Soc.* **2015**, *162*, A5054. [CrossRef]
26. Frackowiak, E.; Meller, M.; Menzel, J.; Gastol, D.; Fic, K. Redox-active electrolyte for supercapacitor application. *Faraday Discuss.* **2014**, *172*, 179–198. [CrossRef]
27. Zhang, L.; Yang, S.; Chang, J.; Zhao, D.; Wang, J.; Yang, C.; Cao, B. A review of redox electrolytes for supercapacitors. *Front. Chem.* **2020**, *8*, 413. [CrossRef]
28. Qin, W.; Zhou, N.; Wu, C.; Xie, M.; Sun, H.; Guo, Y.; Pan, L. Mini-review on the redox additives in aqueous electrolyte for high performance supercapacitor. *ACS Omega* **2020**, *5*, 3801–3808. [CrossRef]
29. Bataev, I.; Panagiotopoulos, N.T.; Charlot, F.; Jorge Junior, A.M.; Pons, M.; Evangelakis, G.A.; Yavari, A.R. Structure and deformation behavior of Zr–Cu thin films deposited on Kapton substrates. *Surf. Coat. Technol.* **2014**, *239*, 171–176. [CrossRef]
30. Siburian, R.; Sihotang, H.; Raja, S.L.; Supeno, M.; Simanjuntak, C. New route to synthesize of graphene nano sheets. *Orient. J. Chem.* **2018**, *34*, 182–187. [CrossRef]
31. Luo, X.-F.; Yang, C.-H.; Peng, Y.-Y.; Pu, N.-W.; Ger, M.-D.; Hsieh, C.-T.; Chang, J.-K. Graphene nanosheets, carbon nanotubes, graphite, and activated carbon as anode materials for sodium-ion batteries. *J. Mater. Chem. A* **2015**, *3*, 10320–10326. [CrossRef]
32. Rosher, S.; Hoffmann, R.; Ambacher, O. Determination of the graphene–graphite ratio of graphene powder by Raman 2D band symmetry analysis. *Anal. Methods* **2019**, *11*, 1224–1228. [CrossRef]
33. Chong, C.; Guo, H.; Zhang, Y.; Hu, Y.; Zhang, Y. Raman study of Sstrain relaxation from grain boundaries in epitaxial graphene grown by chemical vapor deposition on SiC. *Nanomaterials* **2019**, *9*, 372. [CrossRef] [PubMed]
34. Wu, J.-B.; Lin, M.-L.; Cong, X.; Liu, H.-N.; Tan, P.-H. Raman spectroscopy of graphene-based materials and its applications in related devices. *Chem. Soc. Rev.* **2018**, *47*, 1822–1873. [CrossRef] [PubMed]
35. Das, A.; Chakraborty, B.; Sood, A.K. Raman spectroscopy of graphene on different substrates and influence of defects. *Bull. Mater. Sci.* **2008**, *31*, 579–584. [CrossRef]
36. Stubrov, Y.; Nikolenko, A.; Gubanov, V.; Strelchuk, V. Manifestation of structure of electron bands in double-resonant raman spectra of single-walled carbon nanotubes. *Nanoscale Res. Lett.* **2016**, *11*, 2. [CrossRef] [PubMed]
37. Sultana, T.; Georgiev, G.L.; Auner, G.; Newaz, G.; Herfurth, H.J.; Ratwa, R. XPS analysis of laser transmission micro-joint between poly (vinylidene fluoride) and titanium. *Appl. Surf. Sci.* **2008**, *255*, 2569–2573. [CrossRef]
38. Morgan, D.J. Comments on the XPS analysis of carbon materials. *C* **2021**, *7*, 51. [CrossRef]
39. Greczynski, G.; Hultman, L. X-ray photoelectron spectroscopy: Towards reliable binding energy referencing. *Prog. Mater. Sci.* **2020**, *107*, 100591. [CrossRef]
40. Leofanti, G.; Padovan, M.; Tozzola, G.; Benturelli, B. Surface area and pore texture of catalysts. *Catal. Today* **1998**, *41*, 207–219. [CrossRef]
41. Senthikumar, S.T.; Selvan, R.K.; Melo, J.S. Redox additive/active electrolytes: A novel approach to enhance the performance of supercapacitors. *J. Mater. Chem. A* **2013**, *1*, 12386–12394. [CrossRef]
42. Park, Y.; Choi, H.; Lee, D.-G.; Kim, M.-C.; Tran, N.A.T.; Cho, Y.; Lee, Y.-W.; Sohn, J.I. Rational design of electrochemical iodine-based redox mediators for water-proofed flexible fiber supercapacitors. *ACS Sustain. Chem. Eng.* **2020**, *8*, 2409–2415. [CrossRef]
43. Choi, H.; Kim, M.-C.; Park, Y.; Lee, S.; Ahn, W.; Hong, J.; Soh, J.I.; Jang, A.-R.; Lee, Y.-W. Electrochemically active hydroquinone-based redox mediator for flexible energy storage system with improved charge storing ability. *J. Colloid Interface Sci.* **2021**, *588*, 62–69. [CrossRef] [PubMed]
44. Park, Y.; Choi, H.; Kim, M.-C.; Tran, N.A.T.; Cho, Y.; Sohn, J.I.; Hong, J.; Lee, Y.-W. Effect of ionic conductivity in polymer-gel electrolytes containing iodine-based redox mediators for efficient, flexible energy storage systems. *J. Ind. Eng. Chem.* **2021**, *94*, 384–389. [CrossRef]
45. Boota, M.; Hatzell, K.B.; Kumbur, E.C.; Gogotsi, Y. Towards high-energy-density pseudocapacitive flowable electrodes by the incorporation of hydroquinone. *ChemSusChem* **2015**, *8*, 835–843. [CrossRef]
46. Geng, C.-L.; Fan, L.-Q.; Wang, C.-Y.; Wang, Y.-L.; Sun, S.-J.; Song, Z.-Y.; Liu, N.; Wu, J.-H. High energy density and high working voltage of a quasi-solid-state supercapacitor with a redox-active ionic liquid added gel polymer electrolyte. *New J. Chem.* **2019**, *43*, 18935–18942. [CrossRef]
47. Jinisha, B.; Anilkumar, K.M.; Manoj, M.; Ashraf, C.M.; Pradeep, V.S.; Jayalekshmi, S. Solid-state supercapacitor with impressive performance characteristics, assembled using redox-mediated gel polymer electrolyte. *J. Solid State Electrochem.* **2019**, *23*, 3343–3353. [CrossRef]
48. Wada, H.; Yoshikawa, K.; Nohara, S.; Furukawa, N.; Inoue, H.; Sugoh, N.; Iwasaki, H.; Iwakura, C. Electrochemical characteristics of new electric double layer capacitor with acidic polymer hydrogel electrolyte. *J. Power Source* **2006**, *159*, 1464–1467. [CrossRef]
49. Mourad, E.; Coustan, L.; Lannelongue, P.; Zigah, D.; Mehdi, A.; Vioux, A.; Freunberger, S.A.; Favier, F.; Fontaine, O. Biredox ionic liquids with solid-like redox density in the liquid state for high-energy supercapacitors. *Nat. Mater.* **2017**, *16*, 446–453. [CrossRef]
50. Yu, F.; Huang, M.; Wu, J.; Qiu, z.; Fan, L.; Lin, J.; Lin, Y. A redox-mediator-doped gel polymer electrolyte applied in quasi-solid-state supercapacitors. *J. Appl. Polym. Sci.* **2014**, *131*, 39784. [CrossRef]

Article

Vanadium Pentoxide Nanofibers/Carbon Nanotubes Hybrid Film for High-Performance Aqueous Zinc-Ion Batteries

Xianyu Liu [1], Liwen Ma [2], Yehong Du [2], Qiongqiong Lu [3,*], Aikai Yang [4] and Xinyu Wang [2,*]

[1] School of Chemistry and Chemical Engineering, Lanzhou City University, Lanzhou 730070, China; xyliu15@mail.ustc.edu.cn
[2] Institute of Materials and Technology, Dalian Maritime University, Dalian 116026, China; maliwen2019@163.com (L.M.); duyhdlmu@foxmail.com (Y.D.)
[3] Leibniz Institute for Solid State and Materials Research (IFW) Dresden e.V., Helmholtzstraße 20, 01069 Dresden, Germany
[4] Forschungszentrum Jülich GmbH, Institute of Energy and Climate Research, Materials Synthesis and Processing (IEK-1), 52425 Jülich, Germany; a.yang@fz-juelich.de
* Correspondence: q.lu@ifw-dresden.de (Q.L.); wangxinyu@dlmu.edu.cn (X.W.)

Abstract: Aqueous zinc-ion batteries (ZIBs) with the characteristics of low production costs and good safety have been regarded as ideal candidates for large-scale energy storage applications. However, the nonconductive and non-redox active polymer used as the binder in the traditional preparation of electrodes hinders the exposure of active sites and limits the diffusion of ions, compromising the energy density of the electrode in ZIBs. Herein, we fabricated vanadium pentoxide nanofibers/carbon nanotubes (V_2O_5/CNTs) hybrid films as binder-free cathodes for ZIBs. High ionic conductivity and electronic conductivity were enabled in the V_2O_5/CNTs film due to the porous structure of the film and the introduction of carbon nanotubes with high electronic conductivity. As a result, the batteries based on the V_2O_5/CNTs film exhibited a higher capacity of 390 mAh g^{-1} at 1 A g^{-1}, as compared to batteries based on V_2O_5 (263 mAh g^{-1}). Even at 5 A g^{-1}, the battery based on the V_2O_5/CNTs film maintained a capacity of 250 mAh g^{-1} after 2000 cycles with a capacity retention of 94%. In addition, the V_2O_5/CNTs film electrode also showed a high energy/power density (e.g., 67 kW kg^{-1}/ 267 Wh kg^{-1}). The capacitance response and rapid diffusion coefficient of Zn^{2+} (~10^{-8} cm^{-2} s^{-1}) can explain the excellent rate capability of V_2O_5/CNTs. The vanadium pentoxide nanofibers/carbon nanotubes hybrid film as binder-free cathodes showed a high capability and a stable cyclability, demonstrating that it is highly promising for large-scale energy storage applications.

Keywords: aqueous zinc-ion battery; vanadium pentoxide; carbon nanotubes; hybrid film

1. Introduction

The lithium-ion battery is widely used in daily life owing to its many advantages including a high operating voltage, high specific capacity, and long cycle life [1,2]. However, lithium resources on the earth are limited, and the contradiction between its high price and increasing demand is becoming increasingly prominent. In addition, lithium-ion batteries suffer other issues such as high internal resistance, harmful organic electrolytes, and safety hazards [3,4]. These problems restrict their large-scale applications. Rechargeable aqueous batteries have the merits of low production costs, and the electrolyte used is an aqueous electrolyte with high safety. Therefore, it is expected to supplement lithium-ion batteries for new-generation electrochemical energy storage systems [5–8].

Among rechargeable aqueous batteries, aqueous zinc-ion batteries (ZIBs) have attracted more attention due to the high abundance of metal zinc in the earth's resources, low cost, and nontoxicity [9,10]. As zinc metal foil can be directly used as the anode, the development of cathodes of ZIBs have become a research hotspot. The reported cathode materials for ZIBs mainly include manganese compounds, vanadium oxides, Prussian blue, and organic compounds [11–14]. Among these cathode materials, vanadium pentoxide (V_2O_5) has a unique layered structure with a wide range of valence states (from V^{3+} to V^{5+}), which is conducive to the multielectron transfer providing a high specific capacity [15,16]. However, its ion conductivity is low and its diffusion kinetics is slow, resulting in a poor rate performance and unsatisfied cycle performance. Furthermore, the nonconductive and non-redox active polymer was used as the binder in the traditional preparation of electrodes, which hinders the diffusion of zinc ions and compromises the energy density of the electrode [17]. Therefore, in order to avoid using binders, it is important to design a binder-free V_2O_5 electrode.

In this work, V_2O_5 nanofibers/carbon nanotubes (V_2O_5/CNTs) hybrid films were fabricated and employed as the cathode of ZIBs, and the usage of nonconductive and non-redox active binders was avoided. The network structure of V_2O_5/CNTs film is helpful for improving the electronic and ionic conductivity of the electrode. Compared with batteries with binders, the batteries based on the V_2O_5/CNTs film showed a higher specific capacity and a better cycle stability. This work proved that the electrochemical performance of ZIBs can be improved by the application of binder-free electrodes.

2. Materials and Methods

2.1. Preparation of V_2O_5 Nanofibers

First, 0.75 g of NH_4VO_3 (99%, Aladdin) and 1.25 g of P123 (Sigma-Aldrich, St. Louis, MO, USA) were dissolved in 75 mL of water containing 3.75 mL of 2 M of HCl. The mixture was stirred at room temperature for 7 h and then transferred into a Teflon autoclave. After the autoclave was sealed, it was held at 120 °C for 24 h and then cooled to room temperature. The product was washed with deionized water several times and then freeze-dried to obtain V_2O_5 nanofibers.

2.2. Preparation of V_2O_5/CNTs Hybrid Film Electrodes

Here, 20 mg of V_2O_5 and 15 mg of CNTs (length: 0.5–1.5 μm; diameter ~5 nm; Carbon Solutions Inc., Riverside, CA, USA) were dissolved in 40 mL of DMF; then, the mixture was sonicated to form a mixed suspension. The V_2O_5/CNTs film was fabricated by filtration and then dried in an oven at 80 °C.

2.3. Material Characterizations

Scanning electron microscopy (SEM, Supra-55, Zeiss, Oberkochen, Germany) and transmission electron microscopy (TEM, JEOL2100F, JEOL, Tokyo, Japan) were used to investigate the morphology of the samples. The chemical components of the V2O5/CNTs film were confirmed with X-ray photoelectron spectroscopy (XPS, PHI 1600 ESCA, PerkinElmer, Waltham, MA, USA). The structure of the V_2O_5 nanowires and V_2O_5/CNTs film was characterized using X-ray diffraction (XRD, Rigaku D/Max-3A, Rigaku Corporation, Tokyo, Japan). Raman spectra were recorded by a spectrophotometer (Thermo-Fisher Scientific, Waltham, MA, USA).

2.4. Electrochemical Measurements

Stainless-steel CR2032 coin cells were assembled and tested to evaluate the electrochemical performance of the samples. The cells were assembled using a V_2O_5/CNTs composite film as the cathode, filter paper as the separator, Zn foil as the anode, and 3 M of aqueous $Zn(CF_3SO_3)_2$ solution as the electrolyte. Electrochemical impedance spectroscopy (EIS) was performed using a frequency range between 10 mHz and 100 kHz with an AC perturbation signal of 10 mV. Cyclic voltammetry (CV) of the as-assembled battery was conducted at various scan rates (0.2–1.0 mV·s^{-1}). A CHI 660E electrochemical workstation (Shanghai Chenhua, Shanghai, China) was employed to record the CV and EIS results. A CT2001A LAND electrochemical workstation was used to perform the galvanostatic intermittent titration technique (GITT), galvanostatic charge/discharge (GCD), and cyclic performance, within a voltage window of 0.3–1.5 V. All specific capacities reported in this work are based on the cathode mass.

3. Results

The morphology of the as-prepared V_2O_5 was investigated with a transmission electron microscope (TEM) and scanning electron microscope (SEM). The TEM and SEM images reveal that the V_2O_5 had a nanofiber morphology with a diameter of ~18 nm and lengths of several micrometers (Figure 1a,b). After being mixed with CNTs, the V_2O_5 nanofibers were embedded into the network of CNTs (Figure 1c). Furthermore, the V_2O_5/CNTs electrode showed a freestanding structure (inset of Figure 1c). Elemental mappings confirmed that C, O, and V elements were evenly distributed in the V_2O_5/CNTs nanobelts (Figure 1d). XRD and Raman spectroscopy tests were further performed to investigate the V_2O5 nanofibers and V_2O_5/CNTs film. XRD patterns of the V_2O_5 nanofibers and V_2O_5/CNTs film presented typical (001) and (003) peaks (Figure 2a), which fitted well with the layered V_2O_5 (JCPDS no. 40-1296). Peaks of V_4O_7 were also detected, which may be ascribed to the reduction of V_2O_5 by P123 [18]. The Raman spectrum of V_2O_5/CNTs showed the presence of D and G peaks as compared to that of V_2O_5, indicating the presence of CNTs in composite films [19]. The three peaks located at 139, 280, and 983 cm^{-1} are assigned to the V-O vibration in both the V_2O_5/CNTs and V_2O_5 samples (Figure 2b) [20]. In addition, in the XPS survey spectrum, solely C, V, and O elements were detected, confirming the purity of the as-prepared V_2O_5/CNTs sample (Figure 2c). The peak located at 517.5 eV in the V $2p_{1/2}$ spectrum and the peak at 525.2 eV in the V $2p_{3/2}$ spectrum correspond to V^{5+}, and the peak located at 516.8 eV in the V $2p_{1/2}$ spectrum and the peak at 523.7 eV is related to V^{4+} (Figure 2d) [18]. The surface area of V_2O_5/CNTs hybrid films was measured to be 107 m^2 g^{-1}, as shown in Figure 2e.

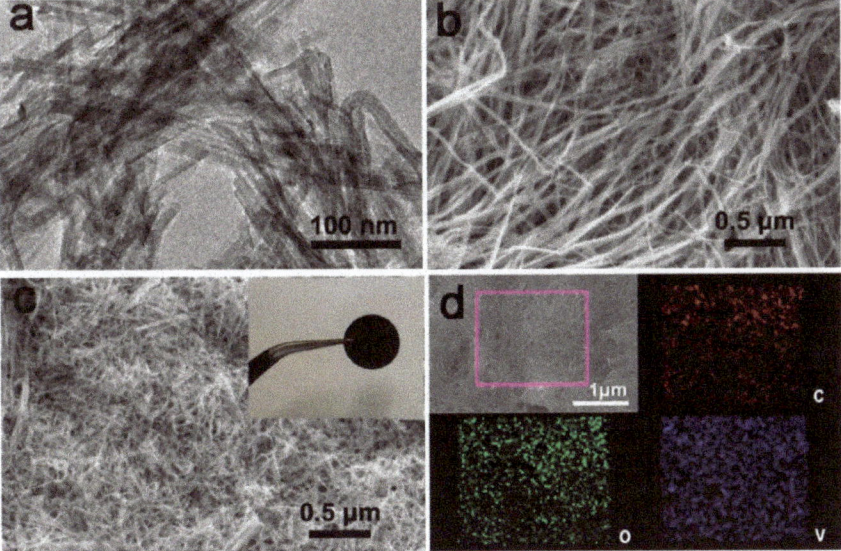

Figure 1. (a) TEM image of the V_2O_5 nanofibers. (b) SEM image of V_2O_5 nanofibers. (c) SEM image and optical image (inset) of V_2O_5/CNTs films. (d) Element mappings of V_2O_5/CNTs.

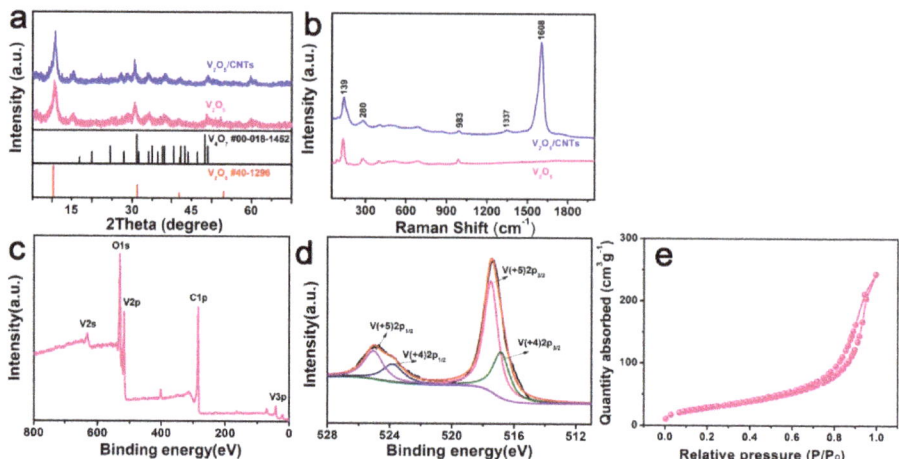

Figure 2. (**a**) XRD patterns of V_2O_5 and V_2O_5/CNTs. (**b**) Raman spectra of the V_2O_5 and V_2O_5/CNTs. (**c**) XPS spectra of V_2O_5/CNTs and (**d**) V 2p spectrum. (**e**) Nitrogen adsorption/desorption isotherms.

The electrochemical properties of V_2O_5 and V_2O_5/CNTs films were further evaluated in ZIBs. The specific capacity at different current densities of V_2O_5 and V_2O_5/CNTs samples are shown in Figure 3a. The V_2O_5/CNTs film showed a high capacity of 399 mAh g^{-1} at 0.1 A g^{-1}, which is higher than that of the V_2O_5 nanofiber (312 mAh g^{-1}). The reason for the capacity decreasing at low current densities is ascribed to the continuous V_2O_5 dissolution [5]. Even at a high current density of 5 A g^{-1}, the V_2O_5/CNTs film still exhibited a high discharge capacity of 239 mAh g^{-1}, while the V_2O_5 nanofiber showed a capacity of 187 mAh g^{-1}. The result demonstrates that the V_2O_5/CNTs film showed a higher rate capability than that of V_2O_5 nanofibers electrode due to the introduction of CNTs. Figure 3b displays the charge/discharge curves of the V_2O_5/CNTs film under various current densities. The charge/discharge curves at different current densities showed similar shapes, indicating the fast charge transfer kinetics of the V_2O_5/CNTs film.

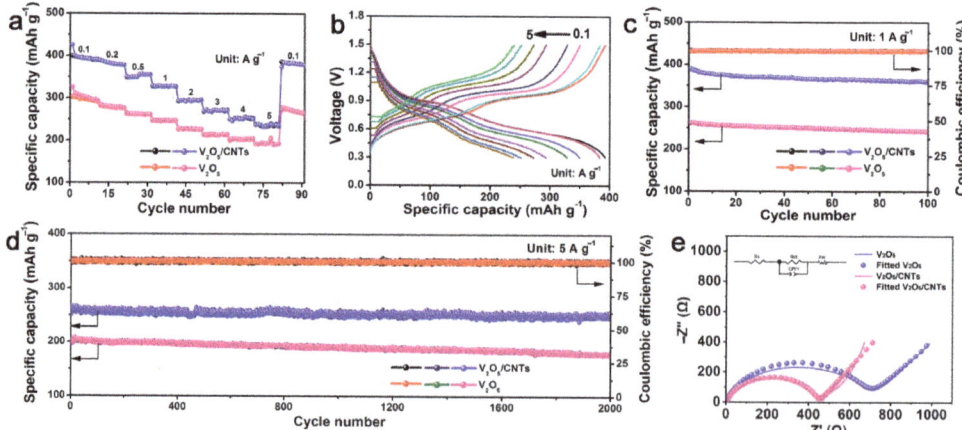

Figure 3. (**a**) The rate performance of the V_2O_5/CNTs film and V_2O_5 electrodes. (**b**) Charge/discharge curves of the V_2O_5/CNTs film and V_2O_5 electrodes at different current densities. (**c**) Cycle performance of V_2O_5/CNTs film and V_2O_5 electrodes. (**d**) Long-term cycling performance of V_2O_5/CNTs film and V_2O_5 electrodes at 5 A g^{-1}. (**e**) Nyquist plots of V_2O_5/CNTs film and V_2O_5 electrodes.

In addition, V_2O_5/CNTs films maintain a high discharge capacity of 273 mAh g^{-1} after 100 cycles at 1 A g^{-1} (Figure 3c). Apart from the good rate capability, the V_2O_5/CNTs film also displayed an excellent long-term cyclic stability. Even at 5 A g^{-1} over 2000 cycles, the batteries based on the V_2O_5/CNTs film maintained a capacity of 251 mAh g^{-1} with a high-capacity retention of 94% (Figure 3d), which is much higher than those of pristine V_2O_5 (168 mAh g^{-1} and 81%). The long cycle capability of the V_2O_5/CNTs film was comparable or higher than most of the previously reported V-based materials without CNTs (Table 1) [21–31]. Furthermore, compared with other works using CNTs in an V_2O_5 electrode, the batteries based on the V_2O_5/CNTs film still displayed a comparable capacity and cycle performance (Table 2) [32–34]. These superior electrochemical performances could be ascribed to the nanowire V_2O_5 knitted with CNTs being helpful for the electrode to keep the close contact and provide an effective electron transmission. The electrochemical impedance spectra (EIS) measurements were performed to study the kinetics. As shown in Figure 3e, both the Nyquist plots of the V_2O_5 and V_2O_5/CNTs film consisted of a hemicycle at the high-frequency region (charge transfer-limited process) and a straight line in the low-frequency region (ion diffusion-limited process). As for the V_2O_5/CNTs sample, the line in the low-frequency region was substantially steeper and the inner diameter of the hemicycle in the high-frequency region was small compared with V_2O_5, manifesting that it had a fast ion diffusion rate and a small resistance. The charge transfer resistance (R_{ct}) of the V_2O_5/CNTs film electrode was about 462 Ω after fitting, which is smaller than that of V_2O_5 (741 Ω), revealing that the introduction of CNTs is beneficial for the high electronic conductivity and efficient Zn^{2+} transport in the V_2O_5/CNTs cathode. Furthermore, the energy/power densities were calculated and compared with other cathode materials (Figure 4). Impressively, the batteries based on the V_2O_5/CNTs film display a remarkable energy density and an impressive power density (e.g., 267 Wh kg^{-1} and 3.2 kW kg^{-1}), which is comparable with the cathodes of $K_2V_6O_{16}\cdot2.7H_2O$, VS_2, $Zn_{0.25}V_2O_5\cdot nH_2O$, LiV_3O_8, $Na_{0.33}V_2O_5$, $Zn_3[Fe(CN)_6]_2$, and $Na_3V_2(PO_4)_3$ [27,30,35–39].

Table 1. The comparison of long-term cycle performances of the V_2O_5/CNTs cathode.

Cathodes	Rate (A g^{-1})	Capacity Retention	Final Capacity (mAh g^{-1})	Reference
V_2O_5/CNTs	5	94% (2000 cycles)	251	This work
$V_2O_5\cdot nH_2O$	6	71.0% (900 cycles)	213	[21]
Cu^{2+}-V_2O_5	10	88.0% (5000 cycles)	180	[22]
K^+-V_2O_5	8	96.0% (1500 cycles)	172	[23]
Graphene/$H_2V_3O_8$	6	87.0% (2000 cycles)	240	[24]
V_2O_5@PANI	5	93.8% (1000 cycles)	201	[25]
2D V_2O_5	20	68.2% (500 cycles)	117	[26]
$Zn_{0.25}V_2O_5\cdot nH_2O$	2.4	80.0% (1000 cycles)	208	[27]
$NaV_3O_8\cdot1.5H_2O$	4	82.0% (1000 cycles)	120	[28]
$Na_2V_6O_{16}\cdot3H_2O$	14	85% (1000 cycles)	129	[29]
$K_2V_6O_{16}\cdot2.7H_2O$	5	88% (229 cycles)	139	[30]
$Na_{1.1}V_3O_{7.9}$/rGO	1	93% (500 cycles)	85	[31]

Table 2. The comparison of the V_2O_5/CNTs cathode with other CNT-based V_2O_5 electrodes.

Cathodes	Specific Capacity	Capacity Retention	Reference
V_2O_5/CNTs	399 mAh g^{-1} (0.1 A g^{-1}) 327 mAh g^{-1} (1 A g^{-1})	5A g^{-1}: 94% (2000 cycles)	This work
V_2O_5/CNTs nanopaper	375 mAh g^{-1} (0.5 A g^{-1})	10A g^{-1}: 80.0% (500 cycles)	[32]
V_2O_5/CNTs (VCP)	312 mAh g^{-1} (1 A g^{-1})	1 A g^{-1}: 81% (2000 cycles)	[33]
V_2O_5@CNTs	293 mAh g^{-1} (0.3 A g^{-1})	5 A g^{-1}: 72.0% (6000 cycles)	[34]

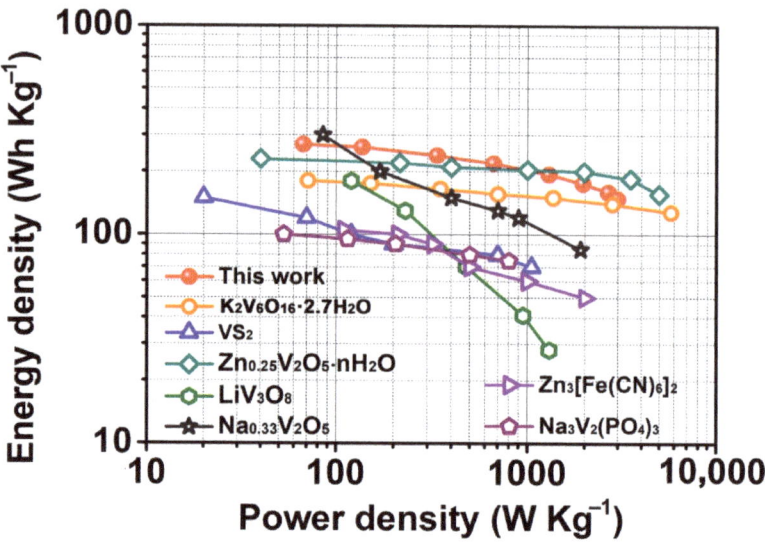

Figure 4. Ragone plot of the batteries based on V$_2$O$_5$/CNTs compared with oher reported data for ZIBs.

The electrochemical kinetics was further investigated to explain the electrochemical performance. The CV curves of the V$_2$O$_5$/CNTs film was measured at different scan rates. As shown in Figure 5a, the CV curves showed similar shapes with the growth of the scan rates, which indicates its good electrochemical reversibility. The characteristic peaks appeared at 0.5/0.7 V, as well as 0.8/1.0 V, reflecting the redox reaction in V$_2$O$_5$/CNTs (Figure 5a) [15,18]. According to the previous literature, the peak current (i) and scan rates (v) have a linear relationship, which can be written as [40]:

$$i = av^b \tag{1}$$

where a and b are adjustable parameters. When b is close to 1, the reaction is a mainly surface-controlled process; when b is near to 0.5, the reaction is dominated by diffusion-controlled behavior. The slope of the peaks of the V$_2$O$_5$/CNTs film is close to 1, which is higher than that of the V$_2$O$_5$ electrode [15,18,22], indicating that the electrochemical process of the V$_2$O$_5$/CNTs is dominated by the pseudocapacitive behavior (Figure 5b). Furthermore, the contribution of pseudocapacitance at different scan rates can be calculated by the following equation: [41]

$$i = k_1 v + k_2 v^{1/2} \tag{2}$$

The current density (i) should be divided into two parts, the pseudocapacity influence ($k_1 v$) and the diffusion-dominant reaction ($k_2 v^{1/2}$). Based on the integration of the CV curve, 66.3% of the total charge storage of the V$_2$O$_5$/CNTs cathode is from the capacitive contribution at 0.5 mV s^{-1} (Figure 5c). The proportions of the capacitive contribution for the V$_2$O$_5$/CNTs cathode are listed in Table 3 (Figure 5d).

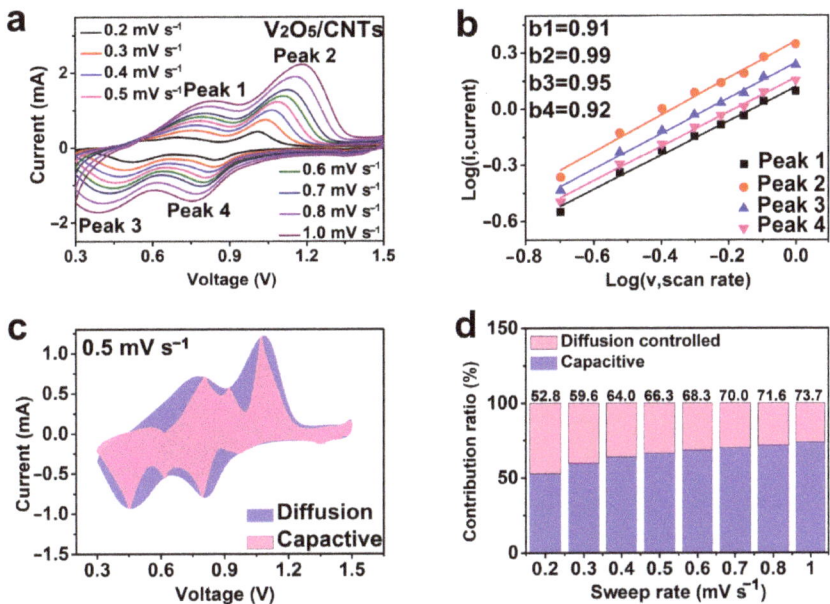

Figure 5. (a) CV curves of the V$_2$O$_5$/CNTs electrode at different scan rates. (b) Log(current) vs. log (scan rate) plots of four peaks in the CV curves. (c) Capacity separation curves at 0.5 mV·s^{-1}. (d) Capacity contribution ratios at multiple scan rates.

Table 3. The proportions of the capacitive contribution for the V$_2$O$_5$/CNTs cathode.

Scan rate (mV s^{-1})	0.2	0.3	0.4	0.5	0.6	0.7	0.8	1.0
Capacitive contribution (%)	52.8	59.6	64.0	66.3	68.3	70.0	71.6	73.7

In order to study the kinetics of Zn^{2+} diffusion in these batteries, a constant-current intermittent titration technique (GITT) test was performed (Figure 6a). The diffusion coefficients (D) of Zn^{2+} ions at the discharge process and charge process can be estimated according to the following equation [42]:

$$D = \frac{4}{\pi\tau}\left(\frac{m_B V_M}{M_B S}\right)^2 \left(\frac{\Delta E_s}{\Delta E_\tau}\right)^2 \quad (\tau \ll L^2/D) \quad (3)$$

where τ is the time for an applied galvanostatic current; m_B, M_B, and V_M are the mass, molecular weight, and molar volume, respectively; S is the active surface of the electrode (taken as the geometric area of the electrode); ΔE_s and ΔE_τ are the quasi-equilibrium potential and the change in cell voltage E during the current pulse, respectively; L is the average radius of the material particles. In our case, the D_{Zn} value of the battery with the V$_2$O$_5$/CNTs film electrode is ~10^{-8} cm^{-2} s^{-1}, which is higher than the value of the V$_2$O$_5$ cathode (Figure 6b), which is consistent with the CV results. Due to the network structure of the V$_2$O$_5$/CNTs films, high values of the capacitive contribution and diffusion coefficients of Zn^{2+} are enabled, leading to a high rate capability of V$_2$O$_5$/CNTs films. All the above results conclusively substantiate that V$_2$O$_5$/CNTs possesses a bright future for the practical application of ZIBs.

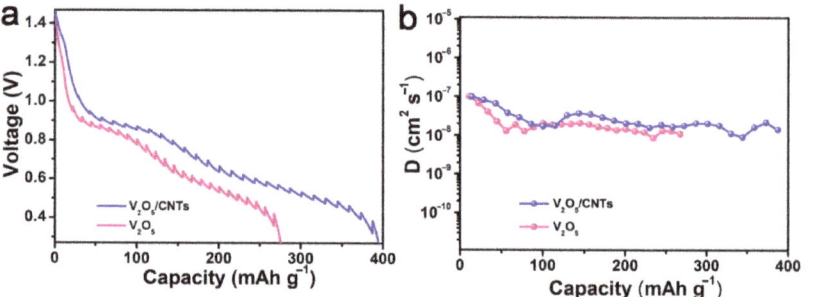

Figure 6. (a) GITT measurements and (b) the corresponding Zn^{2+} diffusion coefficients of V_2O_5/CNTs and V_2O_5 in the discharge process.

4. Conclusions

In summary, V_2O_5/CNTs films were fabricated and employed as binder-free cathodes for ZIBs. The V_2O_5/CNTs film electrodes without nonconductive and non-redox active binders are beneficial for the exposure of active sites and the transfer of electrons and zinc ions, enhancing the electrochemical performance. As a result, the ZIBs based on V_2O_5/CNTs film electrodes possess an excellent rate performance and stable cycle life. This work provides a viable method for fabricating freestanding and binder-free electrodes for energy storage devices and other electronics into highly flexible devices.

Author Contributions: Conceptualization, X.L. and X.W.; methodology, L.M. and X.L.; software, Y.D.; validation, X.L., Q.L. and X.W.; formal analysis, A.Y.; investigation, L.M.; resources, X.W.; data curation, Y.D.; writing—original draft preparation, X.W.; writing—review and editing, Y.D.; visualization, A.Y.; supervision, Q.L.; project administration, X.L.; funding acquisition, X.W. All authors have read and agreed to the published version of the manuscript.

Funding: This work was supported by the Doctoral Research Fund of Lanzhou City University (LZCU-BS2020-03), the Doctoral research startup fund of Liaoning Province (No.2020-BS-066), and the Fundamental Research Funds for the Central Universities (3132019328).

Data Availability Statement: Data is contained within the article.

Conflicts of Interest: The authors declare no conflict of interest.

References

1. Liu, J.; Bao, Z.N.; Cui, Y.; Dufek, E.J.; Goodenough, J.B.; Khalifah, P.; Li, Q.; Liaw, B.Y.; Liu, P.; Manthiram, A.; et al. Pathways for practical high-energy long-cycling lithium metal batteries. *Nat. Energy* **2019**, *4*, 180–186. [CrossRef]
2. Liu, J.Y.; Long, J.W.; Du, S.; Sun, B.; Zhu, S.G.; Li, J.J. Three-Dimensionally porous Li-ion and Li-S battery cathodes: A mini review for preparation methods and energy-storage performance. *Nanomaterials* **2019**, *9*, 441. [CrossRef] [PubMed]
3. Kim, T.; Song, W.T.; Son, D.-Y.; Ono, L.K.; Qi, Y. Lithium-ion batteries: Outlook on present, future, and hybridized technologies. *J. Mater. Chem. A* **2019**, *7*, 2942–2964. [CrossRef]
4. Liu, C.C.; Lu, Q.Q.; Omar, A.; Mikhailova, D. A Facile Chemical Method Enabling Uniform Zn Deposition for Improved Aqueous Zn-Ion Batteries. *Nanomaterials* **2021**, *11*, 764. [CrossRef] [PubMed]
5. Zhang, N.; Chen, X.Y.; Yu, M.; Niu, Z.Q.; Cheng, F.Y.; Chen, J. Materials chemistry for rechargeable zinc-ion batteries. *Chem. Soc. Rev.* **2020**, *49*, 4203–4219. [CrossRef] [PubMed]
6. Ming, J.; Guo, J.; Xia, C.; Wang, W.X.; Alshareef, H.N. Zinc-ion batteries: Materials, mechanisms, and applications. *Mater. Sci. Eng. R* **2019**, *135*, 58–84. [CrossRef]
7. Zhang, N.; Dong, Y.; Wang, Y.Y.; Wang, Y.X.; Li, J.J.; Xu, J.Z.; Liu, Y.C.; Jiao, L.F.; Cheng, F.Y. Ultrafast rechargeable zinc battery based on high-voltage graphite cathode and stable nonaqueous electrolyte. *ACS Appl. Mater. Interfaces* **2019**, *11*, 32978–32986. [CrossRef]
8. Dong, Y.; Jia, M.; Wang, Y.Y.; Xu, J.Z.; Liu, Y.C.; Jiao, L.F.; Zhang, N. Long-life zinc/vanadium pentoxide battery enabled by a concentrated aqueous $ZnSO_4$ electrolyte with proton and zinc ion co-intercalation. *ACS Appl. Energy Mater.* **2020**, *3*, 11183–11192. [CrossRef]

9. Lu, Q.Q.; Liu, C.C.; Du, Y.H.; Wang, X.Y.; Ding, L.; Omar, A.; Mikhailova, D. Uniform Zn Deposition Achieved by Ag Coating for Improved Aqueous Zinc-Ion Batteries. *ACS Appl. Mater. Interfaces* **2021**. [CrossRef]
10. Zhu, M.S.; Hu, J.P.; Lu, Q.Q.; Dong, H.Y.; Karnaushenko, D.D.; Becker, C.; Karnaushenko, D.D.; Li, Y.; Tang, H.M.; Qu, Z.; et al. A patternable and in situ formed polymeric zinc blanket for a reversible zinc anode in a skin-mountable microbattery. *Adv. Mater.* **2021**, *33*, 2007497. [CrossRef]
11. Wang, X.Y.; Qin, X.H.; Lu, Q.Q.; Han, M.M.; Omar, A.; Mikhailova, D. Mixed phase sodium manganese oxide as cathode for enhanced aqueous zinc-ion storage. *Chin. J. Chem. Eng.* **2020**, *28*, 2214–2220. [CrossRef]
12. Wang, X.Y.; Ma, L.W.; Zhang, P.C.; Wang, H.Y.; Li, S.; Ji, S.J.; Wen, Z.S.; Sun, J.C. Vanadium pentoxide nanosheets as cathodes for aqueous zinc-ion batteries with high rate capability and long durability. *Appl. Surf. Sci.* **2020**, *502*, 144207. [CrossRef]
13. Zampardi, G.; La, M.F. Prussian blue analogues as aqueous Zn-ion batteries electrodes: Current challenges and future perspectives. *Curr. Opin. Electrochem.* **2020**, *21*, 84–92. [CrossRef]
14. Glatz, H.; Lizundia, E.; Pacifico, F.; Kundu, D. An organic cathode based dual-ion aqueous zinc battery enabled by a cellulose membrane. *ACS Appl. Energy Mater.* **2019**, *2*, 1288–1294. [CrossRef]
15. Zhang, N.; Dong, Y.; Jia, M.; Bian, X.; Wang, Y.Y.; Qiu, M.D.; Xu, J.Z.; Liu, Y.C.; Jiao, L.F.; Cheng, F.Y. Rechargeable aqueous Zn-V_2O_5 battery with high energy density and long cycle life. *ACS Energy Lett.* **2018**, *3*, 1366–1372. [CrossRef]
16. Wang, X.Y.; Ma, L.W.; Sun, J.C. Vanadium pentoxide nanosheets in-situ spaced with acetylene black as cathodes for high-performance zinc-ion batteries. *ACS Appl. Mater. Interfaces* **2019**, *11*, 41297–41303. [CrossRef]
17. Wan, F.; Zhang, Y.; Zhang, L.L.; Liu, D.B.; Wang, C.D.; Song, L.; Niu, Z.Q.; Chen, J. Reversible oxygen redox chemistry in aqueous zinc-ion batteries. *Angew. Chem. Int. Ed.* **2019**, *58*, 7062–7067. [CrossRef] [PubMed]
18. Qin, X.H.; Wang, X.Y.; Sun, J.C.; Lu, Q.Q.; Omar, A.; Mikhailova, D. Polypyrrole wrapped V_2O_5 nanowires composite for advanced aqueous zinc-ion batteries. *Front. Energy Res.* **2020**, *8*, 199. [CrossRef]
19. Wu, J.B.; Gao, X.; Yu, H.M.; Ding, T.P.; Yan, Y.X.; Yao, B.; Yao, X.; Chen, D.C.; Liu, M.L.; Huang, L. A scalable free-standing V_2O_5/CNT film electrode for supercapacitors with a wide operation voltage (1.6 V) in an aqueous electrolyte. *Adv. Funct. Mater.* **2016**, *26*, 6114–6120. [CrossRef]
20. Jiang, H.F.; Cai, X.Y.; Qian, Y.; Zhang, C.Y.; Zhou, L.J.; Liu, W.L.; Li, B.S.; Lai, L.F.; Huang, W. V_2O_5 embedded in vertically aligned carbon nanotube arrays as free-standing electrodes for flexible supercapacitors. *J. Mater. Chem. A* **2017**, *5*, 23727–23736. [CrossRef]
21. Yan, M.Y.; He, P.; Chen, Y.; Wang, S.Y.; Wei, Q.L.; Zhao, K.N.; Xu, X.; An, Q.Y.; Shuang, Y.; Shao, Y.; et al. Water-lubricated intercalation in $V_2O_5 \cdot nH_2O$ for high-capacity and high-rate aqueous rechargeable zinc batteries. *Adv. Mater.* **2018**, *30*, 1703725. [CrossRef]
22. Yang, Y.Q.; Tang, Y.; Liang, S.Q.; Wu, Z.X.; Fang, G.Z.; Cao, X.X.; Wang, C.; Lin, T.Q.; Pan, A.Q.; Zhou, J. Transition metal ion-preintercalated V_2O_5 as high-performance aqueous zinc-ion battery cathode with broad temperature adaptability. *Nano Energy* **2019**, *61*, 617–625. [CrossRef]
23. Islam, S.; Alfaruqi, M.H.; Putro, D.Y.; Soundharrajan, V.; Sambandam, B.; Jo, J.; Park, S.; Lee, S.; Mathew, V.; Kim, J. K^+ intercalated V_2O_5 nanorods with exposed facets as advanced cathodes for high energy and high rate zinc-ion batteries. *J. Mater. Chem. A* **2019**, *7*, 20335–20347. [CrossRef]
24. Pang, Q.; Sun, C.L.; Yu, Y.H.; Zhao, K.N.; Zhang, Z.Y.; Voyles, P.M.; Chen, G.; Wei, Y.J.; Wang, X.D. $H_2V_3O_8$ Nanowire/Graphene electrodes for aqueous rechargeable zinc ion batteries with high rate capability and large capacity. *Adv. Energy Mater.* **2018**, *8*, 1800144. [CrossRef]
25. Du, Y.H.; Wang, X.Y.; Man, J.Z.; Sun, J.C. A novel organic-inorganic hybrid V_2O_5@polyaniline as high-performance cathode for aqueous zinc-ion batteries. *Mater. Lett.* **2020**, *272*, 127813. [CrossRef]
26. Javed, M.S.; Lei, H.; Wang, Z.L.; Liu, B.T.; Cai, X.; Mai, W.J. 2D V_2O_5 nanosheets as a binder-free high-energy cathode for ultrafast aqueous and flexible Zn-ion batteries. *Nano Energy* **2020**, *70*, 104573. [CrossRef]
27. Kundu, D.; Adams, B.D.; Duffort, V.; Vajargah, S.H.; Nazar, L.F. A high-capacity and long-life aqueous rechargeable zinc battery using a metal oxide intercalation cathode. *Nat. Energy* **2016**, *1*, 16119. [CrossRef]
28. Wan, F.; Zhang, L.L.; Dai, X.; Wang, X.Y.; Niu, Z.Q.; Chen, J. Aqueous rechargeable zinc/sodium vanadate batteries with enhanced performance from simultaneous insertion of dual carriers. *Nat. Commun.* **2018**, *9*, 1656. [CrossRef]
29. Soundharrajan, V.; Sambandam, B.; Kim, S.; Alfaruqi, M.H.; Putro, D.Y.; Jo, J.; Kim, S.; Mathew, V.; Sun, Y.K.; Kim, J. $Na_2V_6O_{16} \cdot 3H_2O$ barnesite nanorod: An open door to display a stable and high energy for aqueous rechargeable Zn-ion batteries as cathodes. *Nano Lett.* **2018**, *18*, 2402–2410. [CrossRef]
30. Sambandam, B.; Soundharrajan, V.; Kim, S.; Alfaruqi, M.H.; Jo, J.; Kim, S.; Mathew, V.; Sun, Y.K.; Kim, J. $K_2V_6O_{16} \cdot 2.7H_2O$ nanorod cathode: An advanced intercalation system for high energy aqueous rechargeable Zn-ion batteries. *J. Mater. Chem. A* **2018**, *6*, 15530–15539. [CrossRef]
31. Cai, Y.S.; Liu, F.; Luo, Z.G.; Fang, G.Z.; Zhou, J.; Pan, A.Q.; Liang, S.Q. Pilotaxitic $Na_{1.1}V_3O_{7.9}$ nanoribbons/graphene as high-performance sodium ion battery and aqueous zinc ion battery cathode. *Energy Storage Mater.* **2018**, *13*, 168–174. [CrossRef]
32. Li, Y.K.; Huang, Z.; Kalambate, P.K.; Zhong, Y.; Huang, Z.; Xie, M.; Shen, Y.; Huang, Y.H. V_2O_5 nanopaper as a cathode material with high capacity and long cycle life for rechargeable aqueous zinc-ion battery. *Nano Energy* **2019**, *60*, 752–759. [CrossRef]
33. Yin, B.; Zhang, S.; Ke, K.; Xiong, T.; Wang, Y.; Lim, B.K.D.; Lee, W.S.V.; Wang, Z.; Xue, J. Binder-free V_2O_5/CNT paper electrode for high rate performance zinc ion battery. *Nanoscale* **2019**, *11*, 19723–19728. [CrossRef] [PubMed]

34. Chen, H.Z.; Qin, H.G.; Chen, L.L.; Wu, J.; Yang, Z.H. V_2O_5@CNTs as cathode of aqueous zinc ion battery with high rate and high stability. *J. Alloys Compd.* **2020**, *842*, 155912. [CrossRef]
35. He, P.; Yan, M.Y.; Zhang, G.B.; Sun, R.M.; Chen, L.N.; An, Q.Y.; Mai, L.Q. Layered VS_2 nanosheet-based aqueous Zn ion battery cathode. *Adv. Energy Mater.* **2017**, *7*, 1601920. [CrossRef]
36. Alfaruqi, M.H.; Mathew, V.; Song, J.; Kim, S.; Islam, S.; Pham, D.T.; Jo, J.; Kim, S.; Baboo, J.P.; Xiu, Z.; et al. Electrochemical zinc intercalation in lithium vanadium oxide: A high-capacity zinc-ion battery cathode. *Chem. Mater.* **2017**, *29*, 1684–1694. [CrossRef]
37. He, P.; Zhang, G.B.; Liao, X.B.; Yan, M.Y.; Xu, X.; An, Q.Y.; Liu, J.; Mai, L.Q. Sodium ion stabilized vanadium oxide nanowire cathode for high-performance zinc-ion batteries. *Adv. Energy Mater.* **2018**, *8*, 1702463. [CrossRef]
38. Zhang, L.Y.; Chen, L.; Zhou, X.F.; Liu, Z.P. Morphology-dependent electrochemical performance of zinc hexacyanoferrate cathode for zinc-ion battery. *Sci. Rep.* **2015**, *5*, 18263. [CrossRef]
39. Li, G.L.; Yang, Z.; Jiang, Y.; Jin, C.H.; Huang, W.; Ding, X.L.; Huang, Y.H. Towards polyvalent ion batteries: A zinc-ion battery based on NASICON structured $Na_3V_2(PO_4)_3$. *Nano Energy* **2016**, *25*, 211–217. [CrossRef]
40. Bin, D.; Huo, W.C.; Yuan, Y.B.; Huang, J.H.; Liu, Y.; Zhang, Y.X.; Dong, F.; Wang, Y.G.; Xia, Y.Y. Organic-inorganic-induced polymer intercalation into layered composites for aqueous zinc-ion battery. *Chem* **2020**, *6*, 968–984. [CrossRef]
41. Du, Y.H.; Wang, X.Y.; Sun, J.C. Tunable oxygen vacancy concentration in vanadium oxide as mass-produced cathode for aqueous zinc-ion batteries. *Nano Res.* **2021**, *14*, 754–761. [CrossRef]
42. Zhang, N.; Jia, M.; Dong, Y.; Wang, Y.Y.; Xu, J.Z.; Liu, Y.C.; Jiao, L.F.; Cheng, F.Y. Hydrated layered vanadium oxide as a highly teversible cathode for rechargeable aqueous zinc batteries. *Adv. Funct. Mater.* **2019**, *29*, 1807331. [CrossRef]

Article

Transparent Heat Shielding Properties of Core-Shell Structured Nanocrystalline Cs$_x$WO$_3$@TiO$_2$

Luomeng Chao, Changwei Sun, Jiaxin Li, Miao Sun, Jia Liu * and Yonghong Ma *

College of Science, Inner Mongolia University of Science and Technology, Baotou 014010, China
* Correspondence: jialiu@imust.edu.cn (J.L.); myh_dlut@126.com (Y.M.)

Abstract: Nanocrystalline tungsten bronze is an excellent near-infrared absorbing material, which has a good potential application in the field of transparent heat shielding materials on windows of automobiles or buildings, but it exhibits serious instability in the actual environment, which hinders its further application. In this paper, we coated the Cs$_x$WO$_3$ nanoparticles with TiO$_2$ to prepare core-shell structured Cs$_x$WO$_3$@TiO$_2$, and its crystal structure and optical properties were studied. The results show that the surface of Cs$_x$WO$_3$ nanoparticles is coated with a layer of TiO$_2$ particles with the size of several nanometers, and the shell thickness can be adjusted by the amount of Ti source. The measurement of optical properties illustrates that TiO$_2$-coated Cs$_x$WO$_3$ exhibits good stability in actual environment, and its transparent heat shielding performance will decrease with the increase in TiO$_2$ shell thickness. This work provides a new route to promote the applications of tungsten bronze as heat shielding materials.

Keywords: tungsten bronzes; nanocrystals; core-shell structure; heat-shielding materials

Citation: Chao, L.; Sun, C.; Li, J.; Sun, M.; Liu, J.; Ma, Y. Transparent Heat Shielding Properties of Core-Shell Structured Nanocrystalline Cs$_x$WO$_3$@TiO$_2$. *Nanomaterials* **2022**, *12*, 2806. https://doi.org/10.3390/nano12162806

Academic Editors: Jung-Inn Sohn and Sangyeon Pak

Received: 13 July 2022
Accepted: 9 August 2022
Published: 16 August 2022

Publisher's Note: MDPI stays neutral with regard to jurisdictional claims in published maps and institutional affiliations.

Copyright: © 2022 by the authors. Licensee MDPI, Basel, Switzerland. This article is an open access article distributed under the terms and conditions of the Creative Commons Attribution (CC BY) license (https://creativecommons.org/licenses/by/4.0/).

1. Introduction

Nearly half of the solar radiation energy comes from near infrared light (NIR) in the range of about 760–2500 nm. Therefore, if the glass of buildings or automobiles can block NIR light while maintaining high transmittance of visible light, it can effectively reduce the room temperature, thus reducing the utilization of air conditioning and achieving the purpose of energy saving and emission reduction. At present, the most common commercial transparent heat shielding glass is Low-E glass or ITO glass [1,2], but the popularity of these materials is not very high because of its complex preparation process and high cost. Therefore, the current research has focused on new transparent heat shielding materials such as rare-earth hexaborides, VO$_2$ or tungsten bronzes [3–11]. All of these materials have good application prospects in the field of transparent heat shielding materials, but there are also some problems. It is difficult to prepare nanocrystalline rare-earth hexaborides and the cost is high; high phase transition temperature of 68 °C and low visible transmittance hinders the further development of VO$_2$; although tungsten bronze has excellent properties, its stability in the practical environment is insufficient.

The chemical formula of tungsten bronzes can be written as M$_x$WO$_3$, where the M represents alkaline earth metal, alkali metal, ammonium or rare earth metal ion. When the M cations are inserted into the whole gap of M$_x$WO$_3$, then x = 1. In the actual environment, M$_x$WO$_3$ is easily oxidized, and M$^+$ escapes from the particle and forms WO$_3$ on the surface, which leads to serious instability of NIR absorption of M$_x$WO$_3$ [12]. A good way to solve this problem is coating M$_x$WO$_3$ nanoparticles with suitable materials. Jin et al. prepared core-shell structured Cs$_x$WO$_3$@SiO$_2$ and Cs$_x$WO$_3$@ZnO nanoparticles and achieved high stability of NIR absorption [13,14]. In the previous theoretical calculation, we found that TiO$_2$-coated Cs$_x$WO$_3$ (CWO) also has good transparent heat shielding properties [15]. Considering the better stability of TiO$_2$, we synthesized core-shell structured nanocrystalline CWO@TiO$_2$ in this work, and the stability and optical behavior of nanoparticles was investigated in detail.

2. Materials and methods

2.1. Preparation of CWO Nanoparticles

A total of 0.7554 g of cesium hydroxide monohydrate (CsOH·H_2O) was added to 300 mL benzyl alcohol (C_7H_8O) solution and stirred for half an hour. Then, an appropriate amount of tungsten hexachloride (WCl_6) was added and the WCl_6 concentration of precursor solution was kept at 0.015 M. After that, the orange precursor solution was heated in an autoclave at 200 °C for 4 h. Finally, the obtained blue powder was washed with alcohol and deionized water several times, and dried in a vacuum at 60 °C for 2 h.

2.2. Preparation of CWO@TiO_2 Nanoparticles

An amount of 3 g CWO powder was added to 400 mL ethanol and dispersed by sonication for 30 min. Then, a different amount (0.2, 0.4 and 0.6 mL) of titanium isopropoxide (TTIP) was added to the solution for forming TiO_2 shell on the surface of CWO particles. After that, deionized water was added to the solution with strong stirring. Then, the mixture was heated in the autoclave at 200 °C for 18 h. Finally, the product was washed several times with deionized water, and dried at 50 °C in vacuum overnight. According to the different amount of TTIP used in the reaction process, the obtained three samples are expressed as CWO@0.2TiO_2, CWO@0.4TiO_2 and CWO@0.6TiO_2, respectively.

2.3. Fabrication Process for CWO@TiO_2 Coated Glass

The heat-shielding glass was prepared by the spin coating method. First, 0.2 g CWO@TiO_2 was dispersed in 20 mL ethanol solution by ultrasonic for 30 min, then 5 g polyvinyl butyral (PVB) resin was added under strong stirring for 20 min to obtain coating slurry. After spin-coating with a centrifugal speed of 2000 rad/min for 40 s, the coated glass was kept at 40 °C for 1 h to remove residual liquid.

2.4. Discrete Dipole Approximation Method

Discrete dipole approximation method (DDA) is an effective way to calculate the far-field and near-field optical responses of nanoparticles with complex refractive indexes and arbitrary geometries [16–19]. In this work, an open-source Fortran-90 code DDSCAT 7.3 applying the DDA method [20] was used to simulate the extinction efficiencies of CWO@TiO_2. The program has given ellipsoid, regular tetrahedron, cuboid, cylinder, hexagonal prism, regular tetrahedron and other structural models. These models can be combined with the dielectric function of the corresponding material to calculate the extinction, absorption and scattering efficiency. Specific parameters such as the effective radius of particles, the number of dipoles and the wavelength range should be set in the program during calculation. In our calculation, a complex dielectric constant of CWO measured by Sato et al. was used to simulate the optical response [21] and the dielectric constants of TiO_2 were obtained from the diel files of DDSCAT 7.3. The calculated wavelength range was 300 nm–2500 nm with 100 steps; The effective radius of the CWO@TiO_2 with different TiO_2 shell thicknesses are set as 55 nm, 60 nm, 65 nm, 70 nm and 80 nm; The dipole ratios are 11:10, 6:5, 13:10, 7:5, 8:5; The refractive index is set to 1.

3. Results and Discussions

The phase composition and crystallographic structure of the samples were confirmed by XRD measurement and results are given in Figure 1. The pure CWO presented hexagonal structure of $Cs_{0.32}WO_3$ (JCPDS 83-1334), and no impurity peaks were observed in the pattern. For TiO_2-coated CWO nanoparticles, the extra peaks appeared in the pattern which belongs to TiO_2 (JCPDS 21-1272). With the increase in the amount of Ti source, the peak intensity at 25° is obviously increasing, indicating that the content of TiO_2 is increasing. These XRD results indicate that there are both CWO and TiO_2 crystals in the sample. The two structures exist independently and the formation of TiO_2 did not affect the structure of CWO.

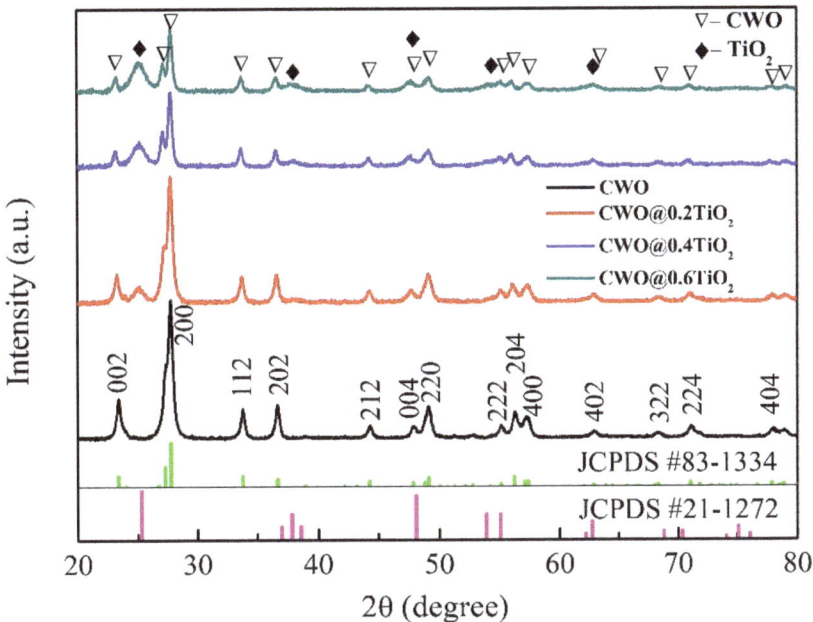

Figure 1. XRD patterns of the uncoated Cs_xWO_3 and TiO_2 coated Cs_xWO_3 nanoparticles.

Figure 2 shows the SEM images of uncoated CWO and TiO_2-coated CWO nanoparticles. The element mapping and corresponding element spectrum for CWO@0.2TiO_2 are also given in Figure 2. The uncoated CWO is composed of irregular particles with good dispersion, and the size is about tens of nanometers. Unlike pure CWO samples, it is clearly seen that a layer of several nanometer-sized small particles appeared on the surface of the CWO particles for the TiO_2 coated samples. With the increase in TiO_2 content, CWO@0.6TiO_2 exhibits obvious spherical shape with largest size. Combined with the XRD results, it can be concluded that CWO is coated with a layer of TiO_2. The element mapping of all samples and corresponding element spectrum for CWO@0.2TiO_2 are also given in Figure 3, which illustrate that Cs, W, O and Ti elements are uniformly distributed in the selected area on coated samples, and no other elements are found.

Figure 4 shows the TEM images of coated CWO samples. It can be clearly seen in Figure 4a–c that tens of nanometers of particles are coated by several nanometers of small particles, and the small particles in the outer layer are increasing with the increase in Ti source. Figure 4d clearly exhibits the single-crystalline nature of the two kinds of particles. The lattice fringes of d = 0.38 nm have good agreement with the (002) crystal planes of CWO structure shown in Figure 1, while the lattice fringe of d = 0.35 nm corresponds with the (101) crystal plane of tetragonal TiO_2 phase. In Figure 4e,f, the diffraction rings of TiO_2 such as (101), (004) and (220) are obtained, which is consistent with the XRD analysis (JCPDS 21-1272). The TEM results confirm that the surface of CWO is coated with a layer of crystalline TiO_2.

Figure 2. SEM image of the (**a**) CWO, (**b**) CWO@0.2TiO$_2$, (**c**) CWO@0.4TiO$_2$, (**d**) CWO@0.6TiO$_2$ (inset shows a magnification of one segment).

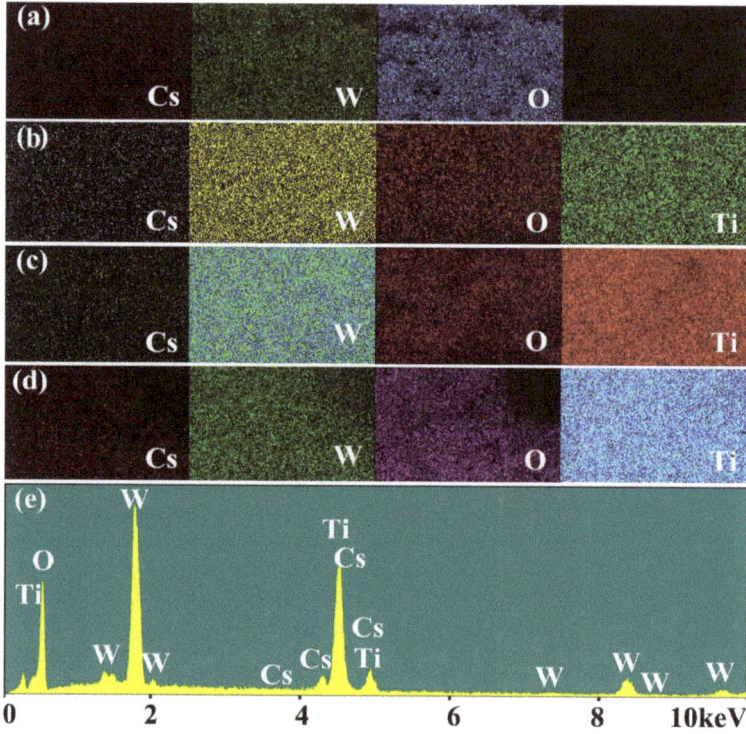

Figure 3. Element mapping of (**a**) CWO, (**b**) CWO@0.2TiO$_2$, (**c**) CWO@0.4TiO$_2$, (**d**) CWO@0.6TiO$_2$; (**e**) Element spectrum of CWO@0.2TiO$_2$.

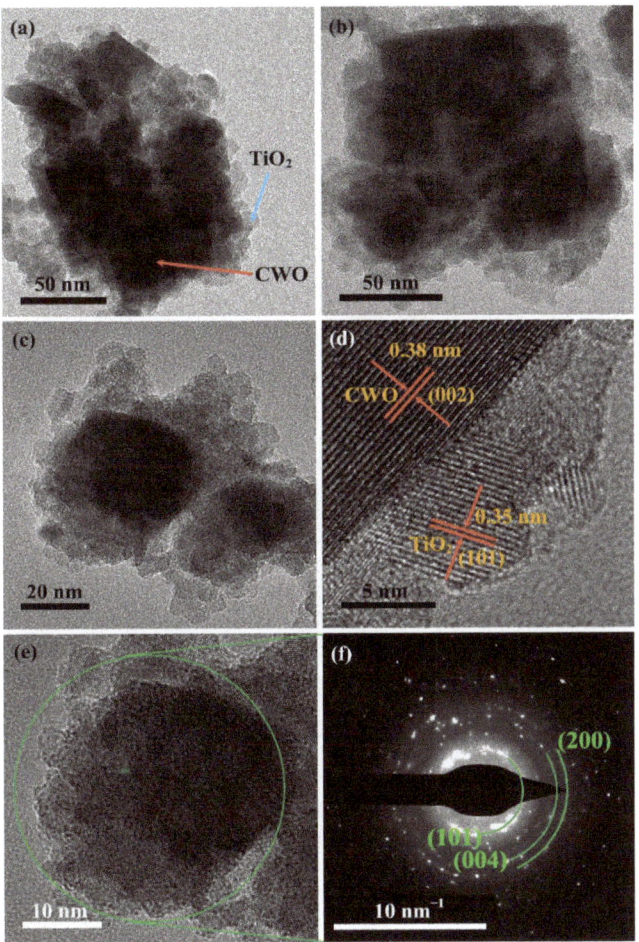

Figure 4. TEM images of the (**a**) CWO@0.2TiO$_2$, (**b**) CWO@0.4TiO$_2$, (**c**) CWO@0.6TiO$_2$; HRTEM image of the (**d**) CWO@0.2TiO$_2$; SAED of the CWO@0.2TiO$_2$ (**e**) selected area, (**f**) diffraction rings.

The chemical states of the coated samples was carefully determined by XPS. The full range XPS spectra and W$_{4f}$ core-level spectra of CWO@0.2TiO$_2$ are given in Figure 5. Except the existence of Cs, W, O, and Ti elements, no other impurity elements were found in the full range XPS spectra, which is consistent with the element spectrum results in Figure 3e. The W$_{4f}$ core-level XPS spectra of CWO@0.2TiO$_2$ can be fitted as two spin-orbit doublets. The peaks at 37.3 and 35.2 eV were attributed to W^{6+}, and the peaks at 36.1 and 33.8 eV were attributed to W^{5+}, respectively. The NIR shielding properties of CWO are determined by the plasmon resonance of free electrons. The addition of Cs element into WO$_3$ structure can reduce part of W^{6+} to W^{5+}, so as to enhance the carrier concentration and small polaron mechanism [22], which is the reason why CWO material has high NIR shielding performance.

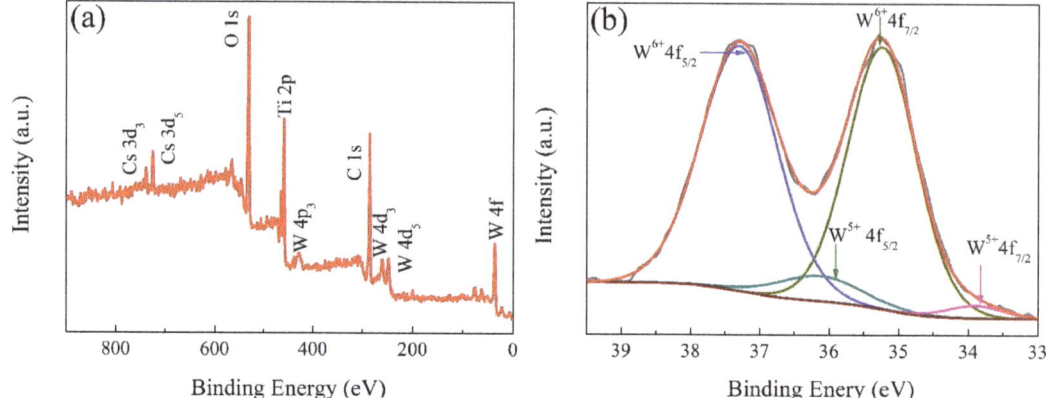

Figure 5. (a) Full range XPS spectra and (b) W_{4f} core-level XPS spectra of CWO@0.2TiO$_2$.

In order to observe the stability of coated CWO in the actual environment, the obtained powders were dispersed in ethanol and made into thin films on glass slides to test their absorption behavior. Figure 6a shows the absorption curve of CWO and CWO@0.2TiO$_2$ nanoparticles after different durations. For the uncoated CWO nanoparticles, the NIR absorption ability degraded seriously after 30 days. While the NIR absorption of CWO@0.2TiO$_2$ nanoparticles showed only slight degradation, indicating that TiO$_2$ showed a good protective effect that TiO$_2$ prevents Cs$^+$ escape from the particle surface and form WO$_3$. To determine the effect of different Ti sources content on the transmittance behavior of CWO particles, the three CWO@TiO$_2$ samples were dispersed with PVB resin and coated on a glass slide (size of 10 cm × 10 cm) using a spin coating method. By observing the CWO@0.2TiO$_2$ coated glass photo in Figure 6b, no opaque or haze-like property was found. The transmittance curve of three CWO@TiO$_2$-coated glass are revealed in Figure 6b, and the transmittance curve of CWO-coated glass is also given for comparison purpose. It can be clearly seen that the four samples show good transparent NIR shielding properties. The transmittance in the visible light decreases with the increase in Ti source and increases obviously in the NIR region. This shows that the thicker the TiO$_2$ shell, the worse the transparent heat shielding performance. This can be attributed to two reasons, one is the influence of the intrinsic optical properties of TiO$_2$, and the other is the increasing particle size with the increase in TiO$_2$ shell thickness. Our previous research shows that the larger the particle size of transparent heat shielding material, the worse the NIR shielding performance [15].

However, it is difficult to accurately control the particle size, shell thickness and other factors in the experiment. In order to systematically study the effect of shell thickness on the optical properties of CWO, we calculated the extinction characteristics of CWO with different TiO$_2$ thickness by using the discrete dipole approximation (DDA) method. Figure 7 presents the extinction efficiencies of CWO spherical particles with different TiO$_2$ shell thicknesses. It is discernible that uncoated CWO shows high extinction in the NIR region and low extinction in the visible region, indicating the transparent heat-shielding properties of CWO materials. With the increase in TiO$_2$ thickness, the absorption edge in the visible region is red-shifted, and the extinction in the NIR region is weakened. We infer that the red shift of the absorption edge in the visible region is related to the intrinsic optical properties of TiO$_2$, while the weakening of the extinction in the NIR region is related to the increase in particle size. Such a trend of optical response with TiO$_2$ shell thickness obtained by DDA calculation is in good agreement with the experimental results in Figure 6b, indicating that the shell thickness should not be too thick when coating CWO with TiO$_2$.

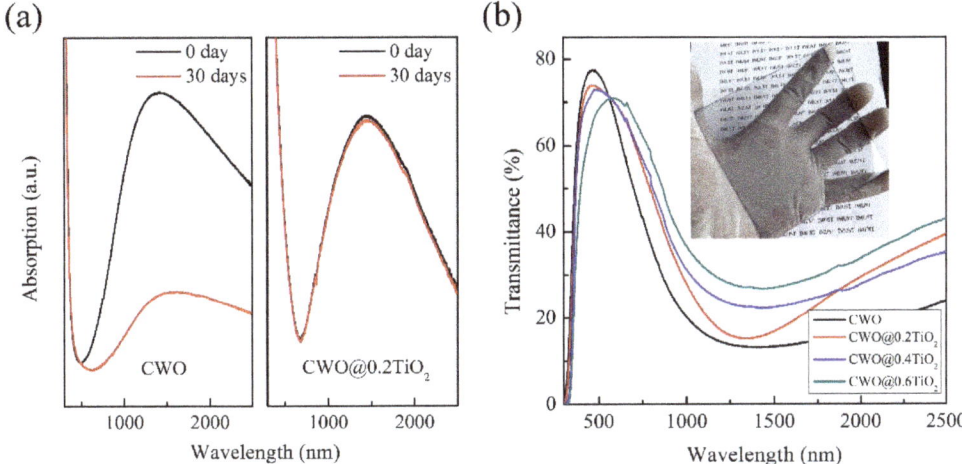

Figure 6. (a) The absorption curve of CWO and CWO@0.2TiO$_2$ nanoparticles after different durations, (b) Transmittance curve of coated glass (inset shows the CWO@0.2TiO$_2$-coated glass used in the transmittance test).

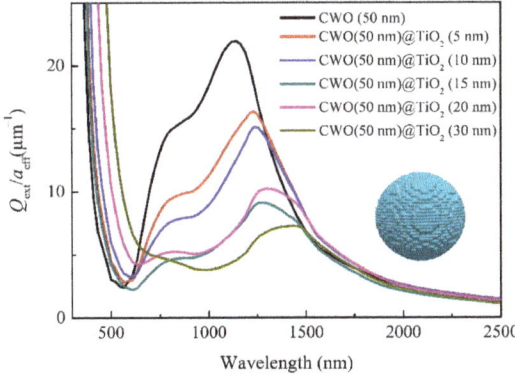

Figure 7. Extinction efficiencies of CWO spherical particles with different TiO$_2$ shell thicknesses.

Finally, to verify the temperature control effect of CWO@TiO$_2$-coated glass, a model house has been designed to test the temperature change as shown in the Figure 8a. The CWO@0.2TiO$_2$-coated glass is placed on the center of a cement wall and directly irradiated by the light from a NIR lamp (PHILIPS, R125) 50 cm away. Two thermocouples of T1 and T2 are placed behind the glass where they are directly illuminated by light and behind the cement wall where not directly illuminated by light, respectively. In addition, a blank glass was used in the control group test. Figure 8b shows the temperature changes with time measured by T1 and T2 in the model house. The results show that the CWO@0.2TiO$_2$ coated glass reduces the T1 and T2 temperature by 6.3 °C and 2.5 °C, respectively, indicating that TiO$_2$-coated CWO still has good heat shielding effect. However, although the other two samples CWO@0.4TiO$_2$ and CWO@0.6TiO$_2$ also have the cooling effect, the effect is not as good as CWO@0.2TiO$_2$, because the thicker the coating, the greater the near-infrared transmittance, which is consistent with the results in Figure 6b.

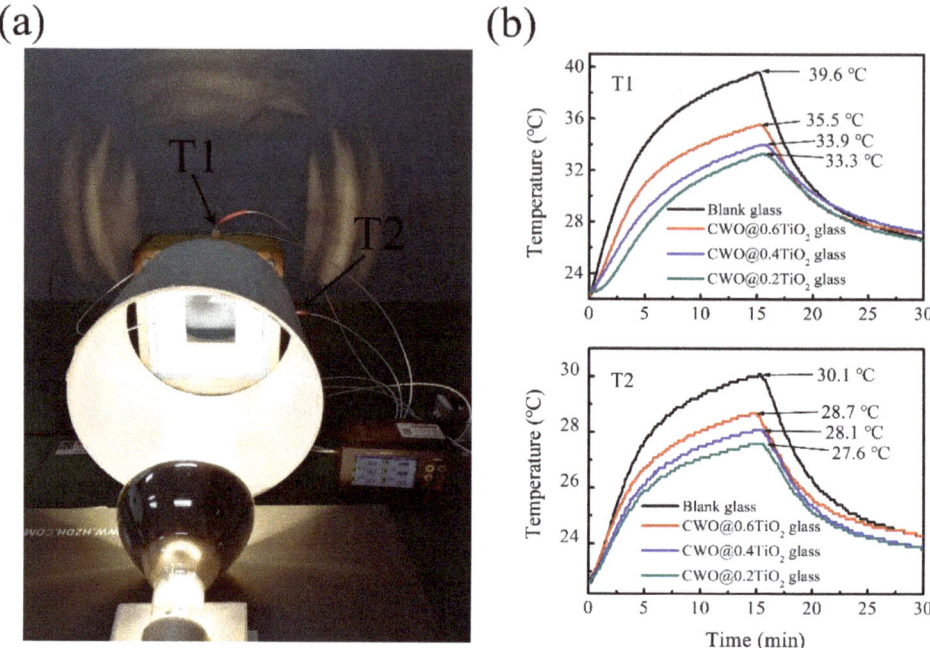

Figure 8. (**a**) Photographs of devices used in temperature control test, (**b**) The temperature changes with time measured by T1 and T2 in the model house.

4. Conclusions

In this article, nanocrystalline CWO particles were prepared by solvothermal method and coated with small TiO_2 crystals. The XRD, SEM and TEM results show that the surface of CWO is coated with a layer of crystalline TiO_2, and the thickness of TiO_2 shell increases with the increase in TTIP amount in the reaction process. The absorption spectrum illustrates that the NIR absorption stability of CWO@0.2TiO_2 is much better than that of CWO after 30 days, indicating that TiO_2 coating significantly improves the stability of tungsten bronze. The transmittance of CWO@TiO_2-coated glass in the visible region decreases with the increase in Ti source and increases obviously in the NIR region, indicating that the thicker the TiO_2 shell, the worse the transparent heat shielding performance. The DDA simulation results also confirm this trend. The measurement of temperature control effect in the model house gives that the CWO@0.2TiO_2-coated glass reduces the indoor temperature by 6.3 °C and 2.5 °C at different places in the room, respectively, which demonstrated the good heat shielding performance of TiO_2-coated CWO.

Author Contributions: L.C.: Resources, Original draft, Supervision, Funding acquisition. C.S.: Investigation, Data curation, Formal analysis. J.L. (Jiaxin Li): Data curation. M.S.: Formal analysis. J.L. (Jia Liu): Supervision. Y.M.: Project administration. All authors have read and agreed to the published version of the manuscript.

Funding: This work was supported by the Natural Science Foundation of Inner Mongolia (No. 2019MS05015), Innovation Fund of Inner Mongolia University of Science and Technology (No. 2019QDL-B35).

Institutional Review Board Statement: Not applicable.

Informed Consent Statement: Not applicable.

Data Availability Statement: No data reported other than that presented.

Conflicts of Interest: The authors declare no conflict of interest. The funders had no role in the design of the study; in the collection, analyses, or interpretation of data; in the writing of the manuscript, or in the decision to publish the results.

References

1. Karlsson, J.; Roos, A. Annual energy window performance vs. glazing thermal emittance—The relevance of very low emittance values. *Thin. Solid. Film.* **2001**, *92*, 345–348. [CrossRef]
2. Kim, C.; Park, J.W.; Kim, J.; Hong, S.J.; Lee, M.J. A highly efficient indium tin oxide nanoparticles (ITO-NPs) transparent heater based on solution-process optimized with oxygen vacancy control. *J. Alloys Compd.* **2017**, *726*, 712–719. [CrossRef]
3. Chao, L.; Bao, L.; Wei, W.; Tegus, O. A review of recent advances in synthesis, characterization and NIR shielding property of nanocrystalline rare-earth hexaborides and tungsten bronzes. *Sol. Energy* **2019**, *190*, 10–27. [CrossRef]
4. Schelm, S.; Smith, G.B. Dilute LaB6 Nanoparticles in Polymer as Optimized Clear Solar Control Glazing. *Appl. Phys. Lett.* **2003**, *82*, 4346. [CrossRef]
5. Takeda, H.; Kuno, H.; Adachi, K. Solar Control Dispersions and Coatings with Rare-Earth Hexaboride Nanoparticles. *J. Am. Ceram. Soc.* **2018**, *91*, 2897. [CrossRef]
6. Long, L.; Ye, H.; Gao, Y.; Zou, R. Performance demonstration and evaluation of the synergetic application of vanadium dioxide glazing and phase change material in passive buildings. *Appl. Energy* **2014**, *136*, 89–97. [CrossRef]
7. Shen, N.; Dong, B.; Cao, C.; Chen, Z.; Liu, J.; Luo, H.; Guo, Y. Lowered phase transition temperature and excellent solar heat shielding properties of well-crystallized VO_2 by W doping. *Phys. Chem. Chem. Phys.* **2016**, *18*, 28010–28017. [CrossRef] [PubMed]
8. Iken, O.; Fertahi, S.; Dlimi, M.; Agounoun, R.; Kadiri, I.; Sbai, K. Thermal and energy performance investigation of a smart double skin facade integrating vanadium dioxide through CFD simulations. *Energy Convers. Manag.* **2019**, *195*, 650–671. [CrossRef]
9. Liang, X.; Guo, C.; Chen, M.; Guo, S.; Zhang, L.; Li, F.; Guo, S.; Yang, H. A roll-to-roll process for multi-responsive soft-matter composite films containing Cs_xWO_3 nanorods for energy-efficient smart window applications. *Nanoscale Horiz.* **2017**, *2*, 319–325. [CrossRef] [PubMed]
10. Ran, S.; Liu, J.; Shi, F.; Fan, C.; Chen, B.; Zhang, H.; Yu, L.; Liu, S.H. Greatly improved heat-shielding performance of K_xWO_3 by trace Pt doping for energy-saving window glass applications. *Sol. Energy. Mater. Sol. Cells* **2018**, *174*, 342–350. [CrossRef]
11. Liu, J.X.; Ando, Y.; Dong, X.L.; Shi, F.; Yin, S.; Adachi, K.; Chonan, T.; Tanaka, A.; Sato, T. Microstructure and electrical–optical properties of cesium tungsten oxides synthesized by solvothermal reaction followed by ammonia annealing. *J. Solid. State. Chem.* **2010**, *183*, 2456–2460. [CrossRef]
12. Adachi, K.; Ota, Y.; Tanaka, H.; Okada, M.; Oshimura, N.; Tofuku, A. Chromatic instabilities in cesium-doped tungsten bronze nanoparticles. *J. Appl. Phys.* **2013**, *114*, 194304. [CrossRef]
13. Chen, Y.; Zeng, X.; Zhou, Y.; Li, R.; Yao, H.L.; Cao, X.; Jin, P. Core-shell structured Cs_xWO_3@ZnO with excellent stability and high performance on near-infrared shielding. *Ceram. Int.* **2018**, *44*, 2738–2744. [CrossRef]
14. Zeng, X.; Zhou, Y.; Jin, S.; Luo, H.; Yao, H.; Huang, X.; Jin, P. The preparation of a high performance nearinfrared shielding Cs_xWO_3/SiO_2 composite resin coating and research on its optical stability under ultraviolet illumination. *J. Mater. Chem. C* **2015**, *3*, 8050–8060. [CrossRef]
15. Sun, C.; Liu, J.; Chao, L. Theoretical analysis on optical properties of $Cs_{0.33}WO_3$ nanoparticles with different sizes, shapes and structures. *Materials. Lett.* **2020**, *272*, 127847. [CrossRef]
16. Draine, B.T. The discrete-dipole approximation and its application to interstellar graphite grains. *Astrophysical. J.* **1988**, *333*, 848–872. [CrossRef]
17. Draine, B.T.; Flatau, P.J. Discrete-dipole approximation for scattering calculations. *J. Opt. Soc. Am.* **1994**, *11*, 1491–1499. [CrossRef]
18. Purcell, E.M.; Pennypacker, C.R. Scattering and absorption of light by nonspherical dielectric grains. *Astrophysical. J.* **1973**, *186*, 705–714. [CrossRef]
19. Yurkin, M.A.; Hoekstra, A.G. The discrete-dipole-approximation code ADDA: Capabilities and known limitations. *J. Quant. Spectrosc. Radiat. Transf.* **2011**, *112*, 2234–2247. [CrossRef]
20. Draine, B.T.; Flatau, P.J. User Guide for the Discrete Dipole Approximation Code DDSCAT 7.3. *arXiv* **2013**. [CrossRef]
21. Sato, Y.; Terauchi, M.; Adachi, K. High energy-resolution electron energy-loss spectroscopy study on the near-infrared scattering mechanism of $Cs_{0.33}WO_3$ crystals and nanoparticles. *J. Appl. Phys.* **2012**, *112*, 074308. [CrossRef]
22. Li, G.; Guo, C.; Yan, M.; Liu, S. CsxWO3, nanorods: Realization of full-spectrum-responsive photocatalytic activities from UV, visible to near-infrared region. *Appl. Catal. B Environ.* **2016**, *183*, 142–148. [CrossRef]

MDPI
St. Alban-Anlage 66
4052 Basel
Switzerland
Tel. +41 61 683 77 34
Fax +41 61 302 89 18
www.mdpi.com

Nanomaterials Editorial Office
E-mail: nanomaterials@mdpi.com
www.mdpi.com/journal/nanomaterials

www.ingramcontent.com/pod-product-compliance
Lightning Source LLC
LaVergne TN
LVHW070600100526
838202LV00012B/519